I0494114

# DISCOVER YOUR OWN DOUBLE STAR

## by

## Martin P Nicholson

**Copyright © 2014 by Martin Nicholson**

All rights reserved. No part of this publication may be reproduced, distributed, or transmitted in any form or by any means, including photocopying, recording, or other electronic or mechanical methods, without the prior written permission of the author, except in the case of brief quotations embodied in critical reviews and certain other noncommercial uses permitted by copyright law. For permission requests email the author at the address below.

Martin Nicholson
Church Stretton
Shropshire SY6 7DQ
United Kingdom

Email – newbinaries@yahoo.co.uk

## Introduction

In my 20 years as an amateur astronomer I have relished working with some of the most gifted, supportive and talented scientists you can imagine. However amateur astronomy in general - and double star astronomy in particular - isn't for the faint hearted or the "shrinking violet" because you will find yourself in the company of a small number of malicious and incompetent "pseudo-scientists" whose destructive influence seems to be out of all proportion to their numbers. If you publish your discoveries or your measurements in a newsgroup, in a peer reviewed (or indeed any) publication or on a blog or a web page they will accuse you of self-aggrandisement but if you chose not to publish your results they will accuse you of low productivity!

> There are many scientifically valid and intellectually stimulating projects that amateur astronomers can carry out in the area of binary and double stars. It is unfortunate that so many people have allowed themselves to be herded down scientific "dead ends" where unnecessarily complicated techniques are used to take measurements that are of negligible scientific value.

If you are serious in your desire to do "proper science" you will need to decide if you want to work alone or as part of a larger team. If you do decide to work with others it is probably sensible to confirm in advance if you are going to get any recognition or thanks for your work and time or whether you are going to be expected to do all the boring and laborious work only to be elbowed out of the way the moment something significant is discovered? It is disappointing that so many professional astronomers still neglect to thank the amateur data collectors when the time comes for peer-reviewed publication.

In 2009 I wrote, *"Both Gänsicke and Shears correctly stress the importance of feedback to the amateur astronomer but I would expand this principle by suggesting that the professionals should actively solicit constructive feedback from their amateur partners. Without mutual feedback and regular communication a parasitic rather than a symbiotic relationship is the almost inevitable result. The professional astronomer should make certain that all the amateur collaborators get full recognition within any published article. Sometimes the amateurs are just grouped together as "members of the AAVSO" or "members of the BAA" with no attempt made to identify individuals. Similarly the professional astronomer should make certain that a copy of the published article is sent to each of the collaborators. It is not uncommon for articles to appear in "subscription only" journals to which the amateur doesn't have access."*

## Why double star astronomy?

There are not many branches of science where amateurs regularly contribute significant observations or discoveries. Astronomy is somewhat different, not least because the number of professional astronomers is relatively small and observing time at the major observatories is very limited. This has left some areas of astronomy where amateurs can contribute. In the past measuring the orientation and separation of binary stars and the identification and characterisation of new binary stars were identified as two examples of "useful" amateur projects.

Two stars orbiting around their joint centre of mass are called binary stars. Binary stars are important in astrophysics for a number of reasons. Orbital studies allow the mass of the stars to be determined. Various sub-types of binary star exist such as optical binaries, spectroscopic binaries and eclipsing binaries. Optical double stars are just "line of sight" arrangements of no scientific importance.

Historically it was very difficult to distinguish between the wheat (binary stars) and the chaff (double stars) and for this reason the Washington Double Star Catalogue (WDS) is a mixture of three types of object.

- Binary stars
- Double stars
- Stars that, without further information, might fall into either class

It is clear from looking at the scientific papers written by amateur astronomers that many of them are still wasting their time measuring systems that the more experienced observers already know or very strongly suspect to be optical double stars. What was needed, what was asked for, but what was never provided, was clear guidance so that amateur energy and enthusiasm could be better targeted.

In December 2006 I wrote: "One of the difficulties that face those of us pushing at the boundary of "standard practice" is the tendency of some colleagues within the field of double star astronomy to interpret any criticism of the status quo as being a criticism of the people involved rather than of the systems currently in use. Instead of a free exchange of ideas the debate usually sinks to the level of name-calling or those in charge simply ignore the questions asked of them."

It is disappointing to have to report that more than 7 years later things have not improved. Indeed in some ways the situation has deteriorated. I wish I was more confident that the few professional astronomers currently working in the area have any genuine desire for a collaborative relationship with amateurs. People say that actions speak louder than words and over the last few years little of substance seems to have been achieved.

**Problem 1** – Currently a small number of systems are seriously over-observed but many others are ignored for decades at a time. Everybody involved in double star astronomy needs to keep reminding themselves that most of the wider pairs of stars only change their appearance very slowly and so they only need to be measured perhaps once a year. There are over 24,000 systems where the discovery announcement is the only entry in the scientific literature and over 50% of all the systems so far reported have been measured 3 or fewer times. One feature of the following table of results requires some explanation. The binary or double star systems with 0 observations have been discovered by the examination of the light curve generated when a star is occulted by an asteroid or the moon.

| Number of observations | Number of stars | Percentage of stars |
|---|---|---|
| 0 | 1229 | 0.96 |
| 1 | 24470 | 19.04 |
| 2 | 19601 | 15.25 |
| 3 | 22341 | 17.38 |
| 4 | 14575 | 11.34 |
| 5 | 9701 | 7.55 |
| 6 or more | 36594 | 28.48 |

Fig. 1 - The number of observations of binary and double star systems in the WDS catalog

| Separation in arc-sec | Number of stars | Number only observed once | Percentage of stars |
|---|---|---|---|
| 0 to 1 | 26829 | 12275 | 45.75 |
| 1 to 2 | 11533 | 4137 | 35.87 |
| 2 to 3 | 8212 | 2538 | 30.91 |
| Over 3 | 81937 | 5520 | 6.74 |

Fig. 2 - Neglected binary and double star systems in the WDS catalogue

It is both difficult and time consuming to measure the separation and position angle of a close double star. A large aperture telescope combined with high magnification and a skilled and experienced observer are all prerequisites and this combination is hard to find.

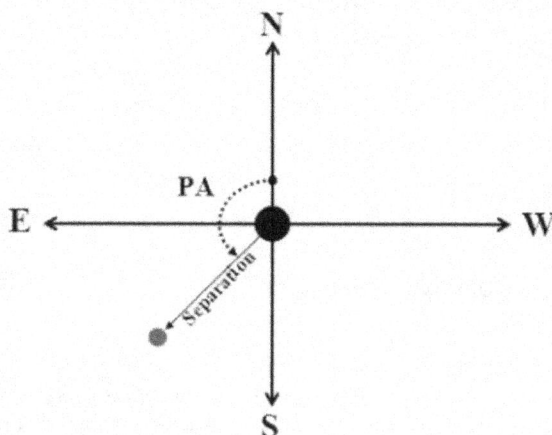

Fig. 3 - The key features of a double star

**Problem 2 -** The system for providing amateur astronomers with feedback on their discoveries or measurements and for then including these results in WDS has some unusual features. Direct submission to professional colleagues is apparently frowned upon and results "have" to be submitted via one of the journals that welcome material of this type. I have found that the requirement that results must appear in a peer reviewed journal as a pre-condition for entry into WDS is applied arbitrarily and inconsistently – there are many examples of results being included following "personal communication" and I know of other examples where peer reviewed results have not appeared at all. It is my belief that there remain some serious conflicts of interest in the current system.

**Problem 3 –** Any astronomer attempting to confirm a discovery claim or wanting to identify possible targets for a research project will require a reliable, easily accessible and up-to-date catalogue that includes all the most recent measurements. What currently exists is an on-line summary of results containing an eclectic mixture of binary stars, double stars and systems that don't exist at all. The problems faced by users are further compounded by the lengthy delay between the making of observations and inclusion in the catalogue. It is not unknown for there to be a three year delay in including peer reviewed results in WDS.

I have found the contrast between the responsive and user-driven systems created by colleagues at the Minor Planet Center and the American Association of Variable Star Observers and those imposed on long-suffering binary star observers rather disconcerting. I worked as a lecturer for many years and I know from experience that rapid feedback and appropriate recognition are strong motivational forces and the systems currently employed by staff at the Washington Observatory fail on both counts.

**Problem 4** – The Washington Double Star Catalogue reports the "discovery code" for each entry. A discovery code is exactly that - it records who carried out the work, the painstaking and time-consuming work, resulting in a new discovery. Discovery codes consist of 1, 2 or 3 letters followed by a number and they have been used in the same way since the earliest days of organised astronomical observation. Any astronomer using the catalogue in 2014 would assume that the discovery code UC meant that the listed object had been discovered by USNO staff – this is certainly not always the case. Discovery codes have never previously been used to identify the source of the raw data used and this new practice described by the Superintendent in his letter to me is, in my opinion, highly questionable ethically and misleading to the wider astronomical community.

**Problem 5** – Over the years I have found a certain lack of clarity, consistency and impartiality from those entrusted with the management and organisation of binary and double star astronomy.

In 2005 I wrote an article entitled "The Daventry Double Star Survey" (J. Br. Astron. Assoc. **115**, 6, 2005). In the article I quoted at length the expert advice I had been given by employees of the United States Naval Observatory (USNO).

Perhaps the most significant piece of advice read as follows:

> What criteria did you use for identifying new doubles? When we have gone through astrometric catalogs (like the AC) looking for new doubles we've only selected ones meeting one of two possible parameters:
>
> – Aitken's parameter: described in the introduction to the ADS, this relates apparent magnitude and separation;
>
> – outer limit: we arbitrarily set this at ten arcseconds, i.e., if the two stars are at 10" or closer we call it a double and put it in the WDS, even if it is so faint that the likely physical separation probably rules out it's being a real double.
>
> If you use criteria similar to this, it might reduce the number of doubles and increase the wheat to chaff ratio.

Fig. 4 - Expert advice from Brian Mason – Project Manager, Washington Double Star Program
Personal communication, December 2002

With hindsight this was a curious statement to have made since the advice makes no mention of the apparent or absolute magnitudes or the spectral class of the stars or their proper motions or the radial velocities of the components.

Nevertheless a few years ago I created a list of candidate double stars based on the written advice that I had been given only to have the entire list rejected. USNO staff then publicly criticised me for using the diagnostic tool that they had suggested to me!

In the years since 2005 there have been many claims of new discoveries based on variations on this "Mason Technique" but few of them seem to have been included in the WDS.

## Projects indended to examine current dilemmas in double star astronomy

**Historical anomalies in the standard catalogue**

It is almost inevitable that any catalogue complied over many years from numerous different sources will contain errors. What is surprising is the lack of enthusiasm shown by amateur and professional astronomers alike to identify and then to rectify these mistakes.

The system ALI 38 was first reported in 1932 as having two components with a separation of 11.5 arc seconds at a position angle of 123 degrees. The quoted magnitudes of 11.5 and 12.6 make this, in theory, a relatively bright double star but there is no sign of the secondary component in this 300 x 300 arc second image from the Digitized Sky Survey. ALI 38 is the exception rather than the rule because the WDS describes the secondary star as a "plate flaw" and the whole system as a "dubious double".

Fig. 5 - ALI 38

POU 160 was first reported in 1898 as having two components (magnitudes 13.1 and 14.2) with a separation of 10.4 arc seconds at a position angle of 142 degrees. This 300 x 300 arc second image, taken in 1990, shows the area around where POU 160 is supposed to be situated. There is one possible candidate on the image but it is a closer and fainter system and a positive match seems unlikely.

Fig. 6 - POU 160

The WDS gives no further information on POU 160 other than labelling it as a "dubious double".

POU 310 is not identified as a "dubious double" in WDS but it certainly doesn't exist at the quoted position. The only report of this system dates back to 1894 (magnitudes 11.35/11.40 at 16 arc seconds and position angle 100 degrees)

Fig. 7 - POU 310

It would be an interesting "rainy day" project to do some detective work on some of the many neglected double stars. For example the systems in the list shown below have only been measured once and that was more than 50 years ago. However they are bright enough and wide enough apart to be fairly easy targets – if they exist!

| RA | DECL | NAME | | DATE | PA | SEP | MAG 1 | MAG 2 |
|---|---|---|---|---|---|---|---|---|
| 00 08 57.10 | +32 57 22.4 | SEI 1 | | 1894 | 232 | 10.6 | 10.5 | 11 |
| 01 33 19.50 | +57 19 35.1 | STI1626 | | 1911 | 48 | 13.1 | 12.1 | 12.1 |
| 02 54 34.44 | -16 22 15.5 | ARA 15 | | 1916 | 5 | 11.9 | 10.88 | 12.9 |
| 03 39 04.47 | -32 50 04.6 | LDS 3520 | | 1960 | 67 | 328 | 12.1 | 12.5 |
| 03 47 14.80 | +25 22 18.5 | POU 310 | | 1894 | 100 | 16 | 11.35 | 11.4 |
| 04 31 44.82 | +15 37 45.8 | LDS 1174 | AC | 1962 | 336 | 12 | 11.9 | 13.4 |
| 05 00 46.40 | +38 10 54.9 | SEI 52 | | 1895 | 215 | 14.4 | 10.5 | 11 |
| 05 28 03.32 | +35 19 33.2 | SEI 266 | AE | 1895 | 266 | 22.7 | 10.8 | 11 |
| 06 10 29.00 | +23 06 54.0 | POU 1101 | | 1905 | 208 | 10.6 | 12.8 | 12.8 |
| 06 22 13.25 | +23 59 52.3 | POU 1265 | | 1925 | 279 | 14 | 11.18 | 12.9 |
| 06 44 08.10 | +31 38 18.5 | SEI 464 | | 1896 | 141 | 24.2 | 11.31 | 11.8 |
| 06 46 43.86 | +23 25 51.3 | POU 2007 | | 1925 | 28 | 12.1 | 11.6 | 13.4 |
| 06 55 01.78 | -01 47 51.1 | BAL 359 | | 1897 | 138 | 16.2 | 11.51 | 11.9 |
| 07 06 36.23 | -02 06 43.6 | BAL 402 | | 1893 | 111 | 17.5 | 11.25 | 11.6 |
| 07 42 31.76 | -18 39 19.6 | ARA 384 | | 1915 | 310 | 14.7 | 11.19 | 11.8 |
| 08 16 48.40 | +33 30 41.0 | SEI 490 | AB | 1894 | 133 | 16 | 11 | 11 |
| 08 16 48.40 | +33 30 41.0 | SEI 491 | AC | 1894 | 141 | 28.1 | 11 | 11 |
| 08 41 53.34 | +35 45 59.1 | SEI 508 | | 1895 | 61 | 25.7 | 11 | 11 |
| 08 49 01.00 | +23 41 18.0 | POU 3010 | | 1898 | 169 | 12.2 | 11.9 | 13.4 |
| 17 12 46.57 | +24 32 40.8 | POU 3264 | | 1892 | 299 | 15.8 | 11.58 | 11.9 |
| 18 08 48.95 | +60 40 41.5 | LDS 2406 | | 1952 | 121 | 30 | 11.1 | 13.4 |
| 19 22 36.95 | +37 08 42.8 | ALI 389 | | 1929 | 61 | 10.3 | 12.22 | 12.8 |
| 19 59 36.98 | +35 09 47.9 | SEI 774 | | 1895 | 105 | 25.4 | 12.53 | 12.67 |
| 20 04 37.03 | +34 55 11.9 | SEI 838 | | 1894 | 222 | 22.5 | 10.5 | 11 |
| 20 05 59.89 | +31 57 33.4 | SEI 852 | | 1893 | 76 | 19 | 11.71 | 11.7 |
| 20 11 48.06 | +31 58 40.2 | SEI 975 | | 1893 | 12 | 11.4 | 11.66 | 12.2 |
| 20 37 17.83 | +03 32 12.3 | BAL2539 | | 1910 | 101 | 18 | 12.16 | 12.3 |
| 20 37 55.47 | +38 05 20.1 | SEI1197 | | 1895 | 174 | 14.7 | 11 | 11 |
| 20 38 13.65 | +38 13 19.5 | SEI1202 | | 1895 | 89 | 14.1 | 11 | 11 |
| 20 47 23.82 | +23 51 35.4 | POU 4989 | | 1899 | 160 | 17.1 | 12.17 | 13.4 |
| 21 26 24.34 | +37 12 02.2 | SEI1514 | | 1895 | 184 | 12 | 11.62 | 12.4 |
| 22 46 24.00 | +23 35 48.0 | POU 5747 | | 1905 | 339 | 15.2 | 12.8 | 13 |
| 23 16 48.98 | +59 26 48.6 | STI2958 | | 1906 | 63 | 10.1 | 10.55 | 10.6 |
| 23 19 26.00 | +24 16 24.0 | POU 5798 | | 1894 | 90 | 19.1 | 11.8 | 11.8 |

Fig. 8 - A selection of neglected double stars

**The quality of the raw data**

At first sight the results shown below look like a particularly good example of a common proper motion binary star system successfully data mined from the Fourth U.S. Naval Observatory CCD Astrograph Catalogue (UCAC4). A pair of fast moving, in astronomical terms, stars with the difference between the proper motion in both right ascension and in declination between the two stars comfortably within the errors in the readings.

| Full | r arcsec | RAJ2000 "h:m:s" | DEJ2000 "d:m:s" | RAJ2000 deg | DEJ2000 deg | f.mag mag | pmRA mas/yr | e (...) | pmDE mas/yr | e (...) | Vmag mag |
|---|---|---|---|---|---|---|---|---|---|---|---|
| 1 | 0.001 | 15 32 01.687 | -40 00 04.33 | 233.0070300 | -40.0012037 | 15.907 | -183.6 | 6.8 | -44.2 | 6.7 | |
| 2 | 96.777 | 15 31 58.128 | -39 58 36.63 | 232.9921989 | -39.9768412 | 14.698 | -182.1 | 7.6 | -31.1 | 8.1 | |

Fig. 9a – The UCAC4 results for a non-existent binary star

If this was just one result out of hundreds contained in an article sent for peer review it is easy to understand how it might escape detailed checking. Both the researcher and the reviewer might be tempted to think, "The information used to identify the pair was obtained from the well-known UCAC4 catalogue so it is bound to be correct."

The reality is that there is no common proper motion binary star system at this location.

A quick check of the relevant area in the Digitized Sky Survey images shows nothing resembling what the catalogue claims to be present. So what has gone wrong?

Fig. 9b - A 5 x 5 arc minute image from the Digitized Sky Survey

A detailed examination of the UCAC4 catalogue entry reveals a number of warning signs. Both the primary and secondary stars have an object classification of 8 – "high proper motion solution in UCAC4, star not matched with PPMXL" and neither star has a matching Two Micron All Sky Survey (2MASS) catalogue entry.

There are over 2 million entries in UCAC4 with this particular object classification - virtually all with a <u>claimed</u> total proper motion of >800 mas/yr. It would be prudent for any potential data-miner to ignore this entire group and also the over 8 million other catalogue entries that have an object classification other than 0.

*However any researcher who examined the UCAC 4 entries with an object classification of 8 to separate the genuine from the spurious would be doing a great service to data miners everywhere.*

**Methodological mishaps**

**Sub group A** – Where the researcher was over optimistic.

GRV 8 is listed in the WDS at a separation of 77.1 arc seconds and a position angle of 302 degrees. Greaves claimed this was a common proper motion pair (Monthly Notices of the Royal Astronomical Society, Volume 355, Issue 2, pp. 585-590) in his 2004 article based on data mining the UCAC2 catalogue.

| Full | r arcsec | RAJ2000 "h:m:s" | DEJ2000 "d:m:s" | RAJ2000 deg | DEJ2000 deg | UCmag mag | pmRA mas/yr | e (...) | pmDE mas/yr | e (...) |
|---|---|---|---|---|---|---|---|---|---|---|
| 1 | 0.012 | 00 08 05.855 | -21 56 34.47 | 002.0243956 | -21.9429087 | 11.24 | 21.9 | 0.8 | -24.1 | 0.6 |
| 2 | 77.711 | 00 08 01.128 | -21 57 15.86 | 002.0047009 | -21.9544045 | 13.51 | 15.8 | 5.6 | -17.6 | 5.7 |

Fig. 10 - The UCAC2 results for the system GRV 8

However the two components have proper motions that differ by 95% (right ascension) and 103% (declination) of the quoted errors and the secondary star also has large percentage errors in the quoted proper motions (35% and 32% respectively). All things considered there was probably not enough evidence in the catalogue to justify the claim that this was a common proper motion pair.

Based on the additional information in the latest version of the catalogue – UCAC4 – it is clear that the system GRV 8 should be deleted. Curiously the staff at the USNO appear unwilling to delete mistakes such as this on the grounds that to do so would be "confusing" for users.

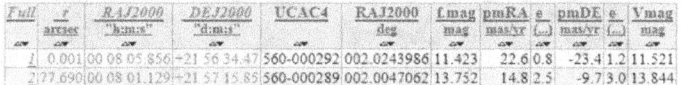

| Full | r arcsec | RAJ2000 "h:m:s" | DEJ2000 "d:m:s" | UCAC4 | RAJ2000 deg | fmag mag | pmRA mas/yr | e (...) | pmDE mas/yr | e (...) | Vmag mag |
|---|---|---|---|---|---|---|---|---|---|---|---|---|
| 1 | 0.001 | 00 08 05.856 | -21 56 34.47 | 560-000292 | 002.0243986 | 11.423 | 22.6 | 0.8 | -23.4 | 1.2 | 11.521 |
| 2 | 77.690 | 00 08 01.129 | -21 57 15.85 | 560-000289 | 002.0047062 | 13.752 | 14.8 | 2.5 | -9.7 | 3.0 | 13.844 |

Fig. 11 The UCAC4 results for the system GRV 8

There is a different problem with GRV 17 which is listed as having a separation of 43.7 arc seconds at a position angle of 215 degrees.

| Full | r arcsec | RAJ2000 "h:m:s" | DEJ2000 "d:m:s" | RAJ2000 deg | DEJ2000 deg | UCmag mag | pmRA mas/yr | e (...) | pmDE mas/yr | e (...) |
|---|---|---|---|---|---|---|---|---|---|---|
| 1 | 0.027 | 00 15 33.012 | +37 55 29.02 | 003.8875500 | +37.9247270 | 12.24 | -15.9 | 0.9 | -13.4 | 1.0 |
| 2 | 43.658 | 00 15 30.882 | +37 54 53.34 | 003.8786739 | +37.9148170 | 13.68 | -17.8 | 4.9 | -16.8 | 4.9 |

Fig. 12 - The UCAC2 results for the system GRV 17

In the UCAC2 catalogue the evidence would support the designation as a common proper motion pair but the revised and improved UCAC4 catalogue strongly suggests that GRV 17 should be removed from WDS because of the large difference in the proper motion in declination for the two components.

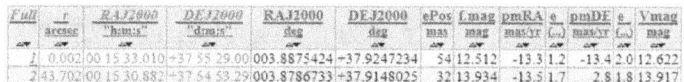

| Full | r arcsec | RAJ2000 "h:m:s" | DEJ2000 "d:m:s" | RAJ2000 deg | DEJ2000 deg | ePos mas | f.mag mag | pmRA mas/yr | e (...) | pmDE mas/yr | e (...) | Vmag mag |
|---|---|---|---|---|---|---|---|---|---|---|---|---|
| 1 | 0.002 | 00 15 33.010 | +37 55 29.00 | 003.8875424 | +37.9247234 | 54 | 12.512 | -13.3 | 1.2 | -13.4 | 2.0 | 12.622 |
| 2 | 43.702 | 00 15 30.882 | +37 54 53.29 | 003.8786733 | +37.9148025 | 32 | 13.934 | -13.5 | 1.7 | 2.8 | 1.8 | 13.917 |

Fig. 13 - The UCAC4 results for the system GRV 17

It is perhaps significant that the latest on-line version of WDS (April 2014) no longer identifies either GRV 8 or GRV 17 as a common proper motion pair.

**Sub group B** – Where the journal editor or the peer reviewer was perhaps over-generous

The April 2014 issue of the Journal of Double Star Observations contains a brief article by Bryant in which he announces the discovery of a new double star in the constellation of Perseus.

http://www.jdso.org/volume10/number2/Bryant_105_106.pdf

TVB 1 was said to have a separation of 14.1 arc seconds at a position angle of 22 degrees. Examination of the Two Micron All Sky Survey (2MASS) data and of the UCAC4 data makes it rather surprising that this pair gained catalogue status.

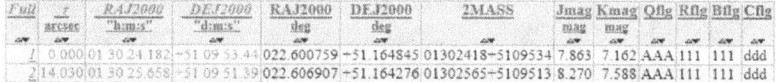

| Full | r arcsec | RAJ2000 "h:m:s" | DEJ2000 "d:m:s" | RAJ2000 deg | DEJ2000 deg | 2MASS | Jmag mag | Kmag mag | Qflg | Rflg | Bflg | Cflg |
|---|---|---|---|---|---|---|---|---|---|---|---|---|
| 1 | 0.000 | 01 30 24.182 | +51 09 53.44 | 022.600759 | +51.164845 | 01302418+5109534 | 7.863 | 7.162 | AAA | 111 | 111 | ddd |
| 2 | 14.030 | 01 30 25.658 | +51 09 51.39 | 022.606907 | +51.164276 | 01302565+5109513 | 8.270 | 7.588 | AAA | 111 | 111 | ddd |

Fig. 14 - 2MASS results for the catalogued pair TVB 1

The stars have similar (J-K) and (B-V) colours so presumably are of similar spectral types but the proper motion results are inconclusive.

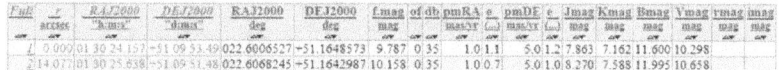

| Full | r arcsec | RAJ2000 "h:m:s" | DEJ2000 "d:m:s" | RAJ2000 deg | DEJ2000 deg | f.mag mag | of db | pmRA mas/yr | e (...) | pmDE mas/yr | e (...) | Jmag mag | Kmag mag | Bmag mag | Vmag mag | r.mag mag | imag mag |
|---|---|---|---|---|---|---|---|---|---|---|---|---|---|---|---|---|---|
| 1 | 0.000 | 01 30 24.157 | +51 09 53.49 | 022.6006527 | +51.1648573 | 9.787 | 0 35 | 1.0 | 1.1 | 5.0 | 1.2 | 7.863 | 7.162 | 11.600 | 10.298 | | |
| 2 | 14.077 | 01 30 25.658 | +51 09 51.48 | 022.6068245 | +51.1642987 | 10.158 | 0 35 | 1.0 | 0.7 | 5.0 | 1.0 | 8.270 | 7.588 | 11.995 | 10.658 | | |

Fig. 15 - UCAC 4 results for TVB 1

TVB 1 strongly reminds me of a pair that I rejected in 2012. I am still not entirely convinced that angular proximity and spectral similarities alone should be enough to secure a catalogue listing.

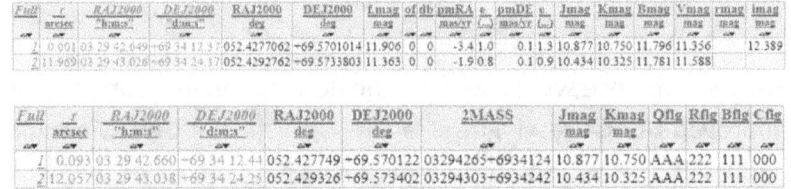

| Full | r arcsec | RAJ2000 "h:m:s" | DEJ2000 "d:m:s" | RAJ2000 deg | DEJ2000 deg | f.mag mag | of | db | pmRA mas/yr | e (...) | pmDE mas/yr | e (...) | Jmag mag | Kmag mag | Bmag mag | Vmag mag | rmag mag | imag mag |
|---|---|---|---|---|---|---|---|---|---|---|---|---|---|---|---|---|---|---|
| 1 | 0.091 | 03 29 42.649 | -69 34 12.37 | 052.4277062 | -69.5701014 | 11.906 | 0 | 0 | -3.4 | 1.0 | | 0.1 | 1.3 | 10.877 | 10.750 | 11.796 | 11.356 | | 12.389 |
| 2 | 11.969 | 03 29 43.026 | -69 34 24.17 | 052.4292762 | -69.5733803 | 11.363 | 0 | 0 | -1.9 | 0.8 | | 0.1 | 0.9 | 10.434 | 10.325 | 11.781 | 11.588 | |  |

| Full | r arcsec | RAJ2000 "h:m:s" | DEJ2000 "d:m:s" | RAJ2000 deg | DEJ2000 deg | 2MASS | Jmag mag | Kmag mag | Qflg | Rflg | Bflg | Cflg |
|---|---|---|---|---|---|---|---|---|---|---|---|---|
| 1 | 0.093 | 03 29 42.660 | -69 34 12.44 | 052.427749 | -69.570122 | 03294265+6934124 | 10.877 | 10.750 | AAA | 222 | 111 | 000 |
| 2 | 12.057 | 03 29 43.038 | -69 34 24.25 | 052.429326 | -69.573402 | 03294303+6934242 | 10.434 | 10.325 | AAA | 222 | 111 | 000 |

Fig. 16a and 16b - Rejected pair 1 – slightly closer but slightly fainter than TVB 1

Fig. 17 - A 5 x 5 arc minute image of my rejected double – perhaps I was rather harsh!

# Data mining for uncatalogued binary and double star systems - FAQ

Q1 - Is there a "one stop shop" where I can look up the details of all the catalogued binary and double stars?

Indeed there is. It is called "The Washington Visual Double Star Catalogue" or WDS for short. You can access WDS at this web address:

http://vizier.hia.nrc.ca/viz-bin/VizieR-3?-source=B/wds/wds

| Full | WDS | Disc | Comp | Obs1 yr | Obs2 yr | Nobs | pa1 deg | pa2 deg | sep1 arcsec | sep2 arcsec | mag1 mag | mag2 mag | pmRA1 mas/yr | pmDE1 mas/yr | pmRA2 mas/yr | pmDE2 mas/yr | Notes |
|---|---|---|---|---|---|---|---|---|---|---|---|---|---|---|---|---|---|
| 1 | 00401+5014 | NI 2 | AB | 1998 | 2002 | 3 | 311 | 311 | 24.9 | 24.80 | 10.68 | 13.00 | 148 | -71 | 143 | -73 | NV |

Fig. 18 - A typical entry in WDS

You will need to build a solid case before any new discovery gets into the catalogue – where a "solid case" means numbers and lots of them.

Q2 - Is it still possible to make a new double star discovery using binoculars?

I would say almost certainly not. A double star that a typical observer could see through binoculars would be brighter than magnitude 8 and with the two components separated by 60 to 200 arc seconds. It is hard to imagine that there are any such double stars that previous generations of dedicated observers have missed.

Q3 - What happens when fainter stars are studied? For example those comfortably observable with a 6 inch aperture telescope at a dark sky site or a CCD equipped telescope in an urban environment?

Read my recent book (May 2014) to find out more.

1800 new double stars for amateur observers
by Martin Nicholson
The book is available on Amazon (www.amazon.com)

This book isn't intended for the astronomical data-miner or the budding astrophysicist so if you buy it expecting to find a detailed analysis of data management techniques or the latest theories on double star formation you will be greatly disappointed. But if you are an amateur astronomer interested in observing double stars this book will fill your heart with joy. If you are tired of seeing the same old lists of double stars with the same, decades old, descriptions then you will love this book because none of these 1800+ double stars have ever been catalogued before! The first few pages contain an introduction the catalogue – the why and the how behind its creation. The heart of the book is a detailed list of these new targets organised by constellation and then by right ascension within the constellation. Detailed information on coordinates, magnitudes, separation, position angle and colour is given for every pair. This book represents a major step forward in popularising double star astronomy.

So to answer the question – yes there are uncatalogued double stars that could be discovered by a modestly equipped amateur astronomer.

Q4 – Is there a quicker way of trying to find uncatalogued binary stars rather than undertaking a laborious "sweeping" of the sky in a project rather reminiscent of comet or nova hunting?

Yes there is - and it is called astronomical data mining.

**Key techniques and terminology in double star data mining.**

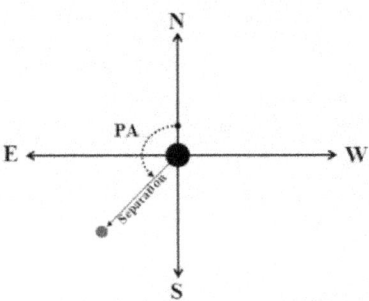

Fig. 19 - The key features of a double star

It is always sensible to look at the characteristics of the entire population before deciding on the fine details of a research project. One of the quickest and easiest ways of doing this is to construct a colour-colour graph. This example was constructed using a random sample of 20,000 stars from SDSS Data Release 10 - each one having clean photometry.

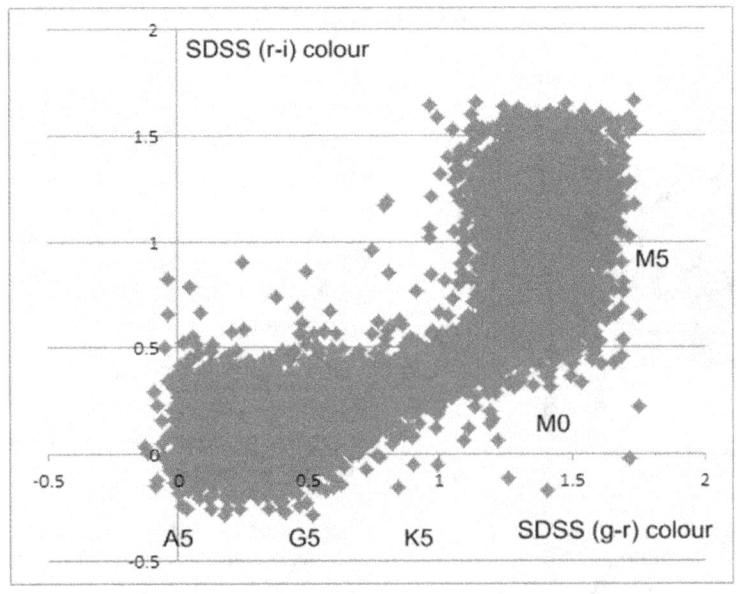

Fig. 20 – A random sample of SDSS stars

Stars of a given spectral class will be found in well-defined areas of the graph. For example red dwarf stars (spectral class M) occupy the top right and the middle right sections of the graph.

| Low galactic extinction | High proper motion |

Fig. 21 – Two different samples of SDSS stars

Changes to the specification of the sample under review can make subtle changes to the shape of the colour-colour graph. The graph for high proper motion stars has a greater percentage of stars with a spectral classification of M. This is not surprising because a high proportion of stars with a high (>60 mas/yr) proper motion that have unsaturated magnitudes in the range covered by SDSS will be red dwarfs. High proper motion suggests that the star is relatively close and a nearby star with an earlier spectral classification - and so intrinsically brighter - will tend to saturate the SDSS sensors.

Careful measurement over many years reveals that all stars are moving independently through space and this causes slow changes in their position relative to the earth. Proper Motion is a vector, with both a magnitude and a direction. The magnitude has units of arc seconds per year and the direction is expressed in degrees with 0 degrees being north, 90 degrees being east and so on. Most catalogues present proper motion information in the form of the magnitude of the motion in both right ascension and in declination since these are at right angles to each other. The third component – radial velocity – is equally valuable as a diagnostic tool but results are not yet available for most stars.

The astronomical data miner will need to be comfortable dealing with large numbers of calculations. Luckily the widespread availability of computer-based spreadsheets has made the identification and characterisation of double stars far quicker and easier than it used to be.

Using RA and Dec values in decimal degrees:

$\Delta_{dec} = (dec_2 - dec_1)$

$\Delta_{RA} = (RA_2 - RA_1) \times \cos(dec_1)$

Then

separation $= \sqrt{\Delta^2_{dec} + \Delta^2_{RA}}$

position angle $= \text{atan}(\Delta_{RA} / \Delta_{dec})$

Fig. 22 - Calculating the separation and position angle between two stars

There are alternatives to using a spreadsheet to convert astrometric and photometric data into double star candidates. The programme "Double Star" can convert the raw data from one million objects into a list of double star candidates in less than an hour.

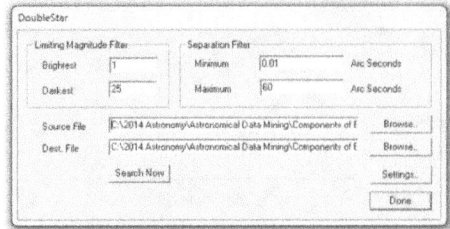

Fig. 23 - The two user selection screens for the software "Double Star"

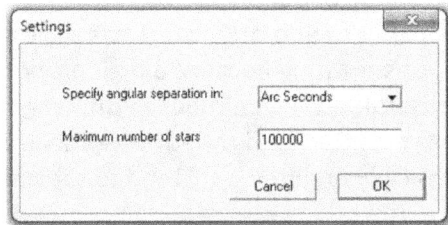

Two particularly useful features of the software – quite apart from the processing speed - are the ability to pre-select both the magnitudes and the range of separation between the two stars.

One of the all-time greatest argument starters in the arcane world of astronomical data miners is, "What constitutes a "discovery" or a "rarity" that is worth reporting?" The author of a research paper almost invariably takes a more liberal approach to what constitutes a new discovery than does the wider astronomical community!

In 2009 I wrote (http://philica.com/display_observation.php?observation_id=55) –

*"In the section on methodology Greaves writes, "… because the purpose was to find as many certain associations as possible, as opposed to any emphasis on quantity." While some of the filters he used in pursuit of this aim are reasonable enough it seems quite extraordinary that anybody would regard two stars with a 50% difference in the quoted values for proper motion in both declination and right ascension as showing common proper-motion."*

A more objective approach to the problem would be to apply the equation suggest by Struve in 1852 whereby the number of observed pairings is compared with the number predicted assuming a random distribution. N is the number of sources matching the selection criteria, A is the survey area in square degrees and $p$ is the maximum separation in degrees. These are combined to calculate n which is the number of pairs that would be expected on a random basis.

$$n(\leqslant \rho) = N(N-1)\pi\rho^2/2A$$

Fig. 24 - Stuve's equation

Using SDSS Data Release 10 (where A = 14555). Suppose 10,000 red dwarf stars match the chosen selection criteria and that these 10,000 stars yield 50 pairs of stars lying within 20 arc seconds (20/3600 of a degree) of each other. The expected number of pairs using the equation shown above would be 0.33 so that the rarity factor is said to be observed/theoretical or 50/0.33 = 150. Few would argue that results of this kind would be worth publishing.

More typical would be the results from an experiment using SDSS Data Release 10 stars that were particularly bright in the u-band (s1.dered_u between 12.1 and 16.1).

```
SELECT TOP 1000000 s1.ra as RA1, s1.dec as DE1, p1.pmra as PM1a, p1.pmdec as PM1b,
s1.psfMag_u as U, s1.psfMag_g as G, s1.psfMag_r as R, s1.psfMag_i as I, s1.psfMag_z as Z,
s1.extinction_u as UE, s1.extinction_g as GE, s1.extinction_r as RE, s1.extinction_i as IE,
s1.extinction_z as ZE

INTO mydb.BrightUStars2
FROM PhotoObjAll S1, ProperMotions p1
WHERE S1.ObjID = p1.ObjID

AND S1.ra between 170 and 190
AND S1.dec between 0 and 20
AND S1.Mode = 1
AND S1.Type = 6
AND S1.Clean = 1

AND s1.dered_u between 12.1 and 16.1
```

When the results were analysed 16 pairs of stars were found with separations of less than 20 arc seconds from the 7,603 stars examined. The expected number of double stars is 7 and the observed number is 16 giving a rarity factor of 2.3 and so these results would be of little astronomical interest.

| # | H | M | S | D | M | S | MAG1 | MAG2 | SEP | PA |
|---|---|---|---|---|---|---|------|------|-----|-----|
| 1 | 12 | 28 | 22.29 | 1 | 50 | 50.41 | 15.509 | 15.945 | 3.78 | 133.56 |
| 2 | 11 | 58 | 50.77 | 12 | 22 | 13.06 | 15.543 | 15.584 | 3.81 | 358.35 |
| 3 | 12 | 39 | 7.7 | 3 | 29 | 43.63 | 14.866 | 16.043 | 5.41 | 353.06 |
| 4 | 12 | 27 | 2.14 | 1 | 32 | 37.19 | 15.05 | 16.066 | 7.52 | 91.77 |
| 5 | 11 | 38 | 42.16 | 11 | 19 | 57.77 | 15.265 | 16.026 | 8.21 | 12.46 |
| 6 | 11 | 40 | 47.1 | 13 | 15 | 54.76 | 15.028 | 15.554 | 8.84 | 116.79 |
| 7 | 11 | 28 | 23.56 | 3 | 49 | 23.85 | 15.268 | 16.102 | 9.48 | 178.08 |
| 8 | 11 | 21 | 23.11 | 12 | 29 | 49.2 | 15.263 | 15.881 | 9.99 | 142.14 |
| 9 | 12 | 13 | 58.7 | 11 | 30 | 37.43 | 15.199 | 15.581 | 11.44 | 201.91 |
| 10 | 11 | 38 | 2.49 | 3 | 19 | 4.67 | 15.108 | 15.264 | 12.74 | 55.04 |
| 11 | 12 | 10 | 4.69 | 13 | 33 | 21.87 | 15.895 | 15.991 | 14.34 | 278.58 |
| 12 | 11 | 56 | 7.65 | 11 | 7 | 16.28 | 15.105 | 16.106 | 16.35 | 60.42 |
| 13 | 11 | 44 | 39.8 | 12 | 14 | 3.95 | 15.144 | 15.389 | 16.84 | 14.88 |
| 14 | 11 | 25 | 49.9 | 1 | 41 | 30.78 | 15.347 | 16.012 | 17.59 | 205.07 |
| 15 | 11 | 27 | 48.59 | 18 | 7 | 54.03 | 15.367 | 15.672 | 19.16 | 325.75 |
| 16 | 11 | 53 | 28.39 | 11 | 46 | 42.91 | 15.73 | 15.761 | 19.36 | 138.94 |

Fig.25 – "bright in the u-band" double stars

# Astronomical data mining of the Sloan Digital Sky Survey – Data Release 10

## Exercise 1 - Misidentifications in data release 10

It isn't generally known that a proportion of the objects identified in the SDSS survey as being galaxies (type 3 in the source classification) are, on examination of the on-line images, seen to be close double stars that were not correctly identified by the software pipeline. As far as the computer software was concerned these objects were non-stellar – due to their elongated shape – and they had the same colour as galaxies or quasars with large redshifts.

For this reason many of these pairs were observed using the multi-object spectrograph providing, serendipitously, far more spectra of red dwarf stars than would otherwise have been the case.

There are three sites that can be used to examine SDSS data. Each has its own advantages and challenges. Currently (May 2014) data release 10 is not available from the Vizier site but data release 9 can be viewed at:-

http://vizier.u-strasbg.fr/viz-bin/VizieR-3?-source=V/139

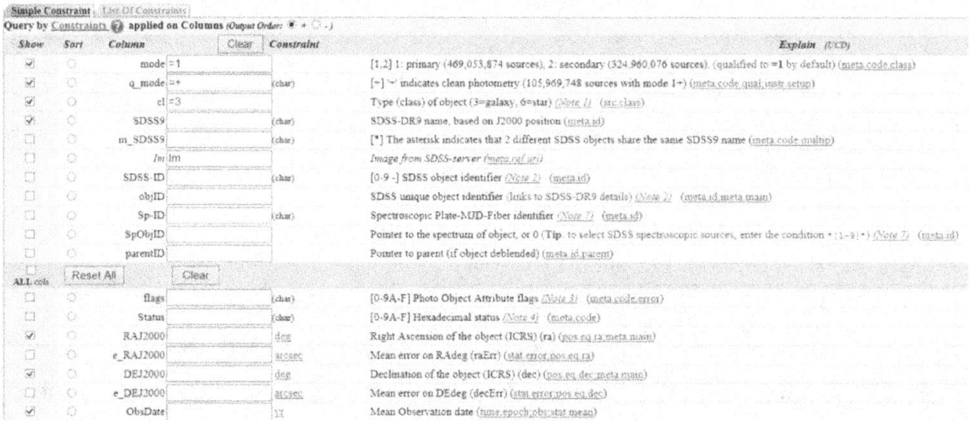

| Show | Sort | Column | Clear | Constraint | Explain (UCD) |
|---|---|---|---|---|---|
| ☑ | ○ | mode | | =1 | [1,2] 1: primary (469,053,874 sources), 2: secondary (324,960,076 sources). (qualified to =1 by default) (meta.code.class) |
| ☑ | ○ | q_mode | | =* (char) | [=] '*' indicates clean photometry (105,969,748 sources with mode 1=) (meta.code.qual;instr.setup) |
| ☑ | ○ | cl | | =3 | Type (class) of object (3=galaxy, 6=star) (Note 2) (src.class) |
| ☑ | ○ | SDSS9 | | (char) | SDSS-DR9 name, based on J2000 position (meta.id) |
| ☐ | ○ | m_SDSS9 | | (char) | [*] The asterisk indicates that 2 different SDSS objects share the same SDSS9 name (meta.code.multip) |
| ☐ | ○ | Im | | Im | Image from SDSS-server (meta.ref.url) |
| ☐ | ○ | SDSS-ID | | (char) | [0-9 -] SDSS object identifier (Note 2) (meta.id) |
| ☐ | ○ | objID | | | SDSS unique object identifier (links to SDSS-DR9 details) (Note 2) (meta.id;meta.main) |
| ☐ | ○ | Sp-ID | | (char) | Spectroscopic Plate-MJD-Fiber identifier (Note 7) (meta.id) |
| ☐ | ○ | SpObjID | | | Pointer to the spectrum of object, or 0 (Tip: to select SDSS spectroscopic sources, enter the condition * (1-9] *) (Note 7) (meta.id) |
| ☐ | ○ | parentID | | | Pointer to parent (if object deblended) (meta.id.parent) |
| ☐ ALL cols | Reset All | | Clear | | |
| ☐ | ○ | flags | | (char) | [0-9A-F] Photo Object Attribute flags (Note 3) (meta.code.error) |
| ☐ | ○ | Status | | (char) | [0-9A-F] Hexadecimal status (Note 4) (meta.code) |
| ☑ | ○ | RAJ2000 | | deg | Right Ascension of the object (ICRS) (ra) (pos.eq.ra;meta.main) |
| ☐ | ○ | e_RAJ2000 | | arcsec | Mean error on RAdeg (raErr) (stat.error;pos.eq.ra) |
| ☑ | ○ | DEJ2000 | | deg | Declination of the object (ICRS) (dec) (pos.eq.dec;meta.main) |
| ☐ | ○ | e_DEJ2000 | | arcsec | Mean error on DEdeg (decErr) (stat.error;pos.eq.dec) |
| ☑ | ○ | ObsDate | | yr | Mean Observation date (time.epoch;obs;stat.mean) |

Fig.26 – The Vizier user interface

The user interface is complex and not ideally suited for searching large areas of the sky and for this type of project it is probably best avoided.

The SDSS Imaging Query Form is much more versatile and more user friendly and can accessed at:-

http://skyserver.sdss3.org/public/en/tools/search/IQS.aspx

However the most useful of the download options (comma separated values or CSV) outputs the results onto the screen rather than giving the user the option to save the results directly to the computer. This is a strange feature and quite unlike how the SDSS treats this format elsewhere on the site.

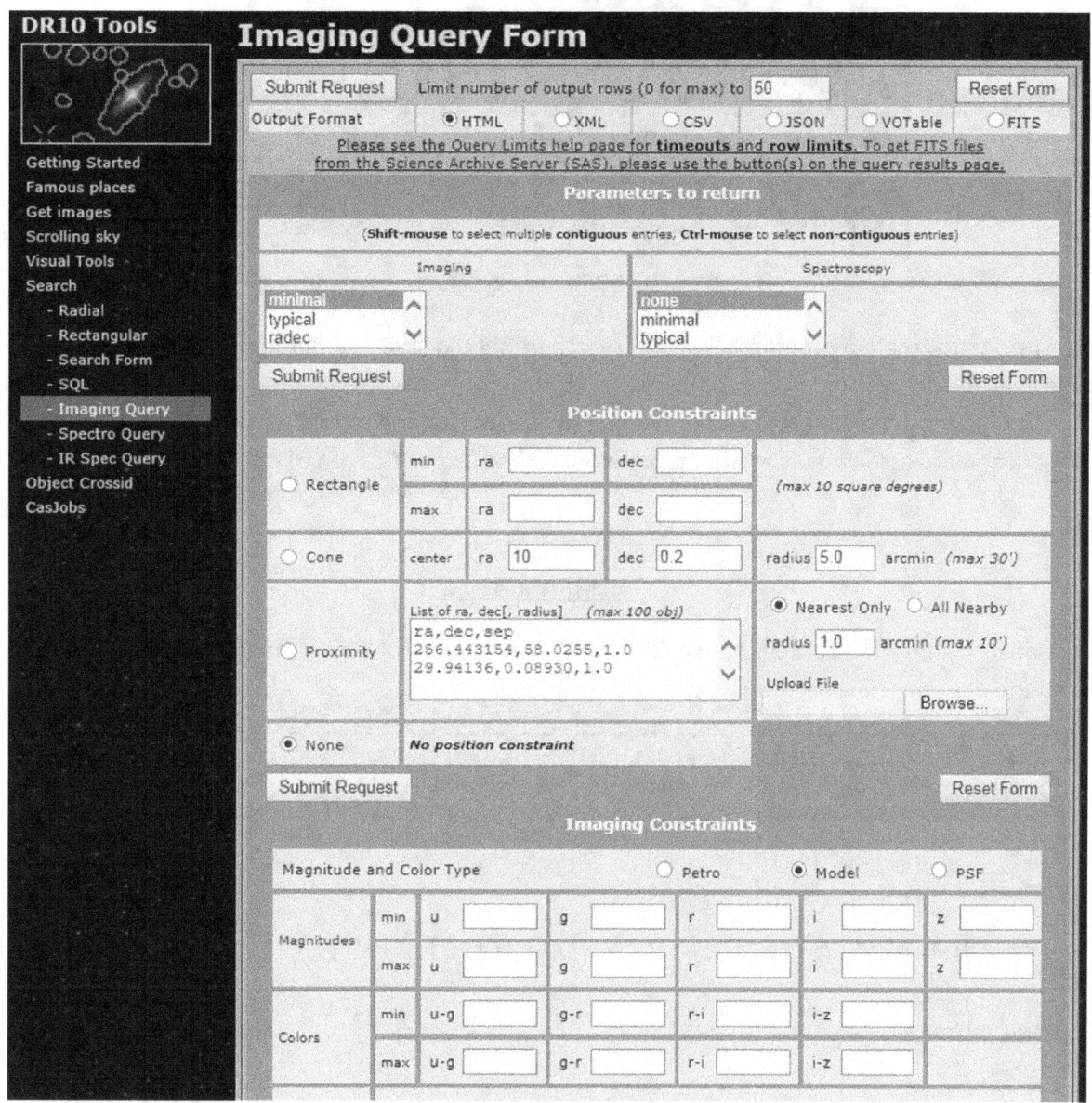

Fig.27 – The SDSS Imaging Query Form

The third – and by far the best option for most data mining projects – is the CAS Jobs page.

http://skyserver.sdss3.org/CasJobs/

Users will need to create an account before they can use this section of the SDSS site but this is a quick and easy process.

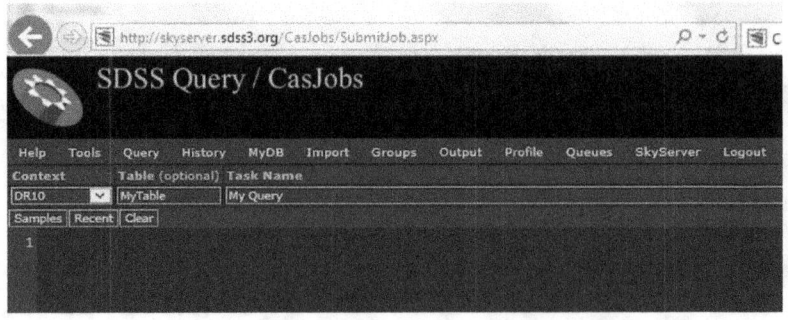

Fig.28 – The SDSS CasJobs Query Form

Before starting to create a program all new researchers should study, in detail, the many sample queries that are available by clicking on the "Samples" button. Once these have been mastered all that needs to be done is to identify the different tables holding the SDSS data and to create the surrounding framework and the associated constraints required to find a specific type of object.

## PROGRAMME 1

It is a relatively simple matter to create a SQL program to search for objects reported as being galaxies but with the photometric characteristics of red dwarf stars.

```
SELECT TOP 5000  s1.ra as RA1, s1.dec as DE1, s1.psfMag_u as U, s1.psfMag_g as G, s1.psfMag_r as R,
s1.psfMag_i as I, s1.psfMag_z as Z, s1.extinction_u as UE, s1.extinction_g as GE, s1.extinction_r as RE,
s1.extinction_i as IE, s1.extinction_z as ZE, p1.pmra as PM1a, p1.pmdec as PM1b

INTO mydb.NotGalaxy1
FROM PhotoObjAll s1, ProperMotions p1

WHERE s1.ObjID = p1.ObjID
AND S1.Mode = 1
AND S1.Type = 3
AND S1.Clean = 1

AND s1.psfMag_i < 16.00

AND s1.psfMag_r between 14.1 and 22.3
AND s1.psfMag_i between 13.8 and 21.7
AND s1.psfMag_z between 12.3 and 20.1
AND s1.extinction_r < 0.1
AND (s1.psfMag_r-s1.psfMag_z) between 0.89 and 4.41
```

Naturally readers are at liberty to modify this program in any way that suits them. I tend to work with a series of pre-prepared modules that I then join together to create a working, rather than a polished, program. Sometimes this technique results in a command becoming superfluous (AND s1.psfMag_z between 12.3 and 20.1 is an example of this).

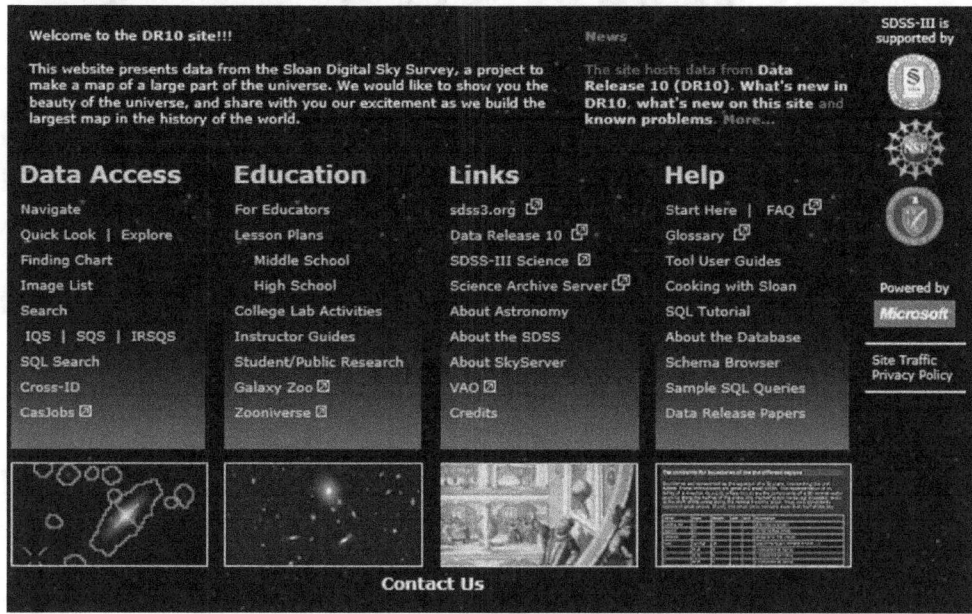

Fig.29 – The SkyServer page that provides links to a wide range of research tools

Once the results have been downloaded it is prudent to examine images of the objects identified by the the SQL programme. The quickest and easiest way to do this is by making use of the "Image List" facility – best accessed via the SkyServer page.

The file that the researcher "cuts and pastes" into the dialog box should contain three columns. A identifying-number column followed by two positional (astrometric) columns.  It is always rather nerve-wracking when the "Get Image" box is selected because it can be at this stage that any problems  with the selection constraints used will first become obvious.

In this experiment I was specifically looking for objects that had not been looked at by the spectrograph. By selecting the "Drawing Option" "Objects with spectra" I was able to identify these objects and eliminate them from further analysis.

The image scale was selected so that close binary stars could be quickly and reliably detected.

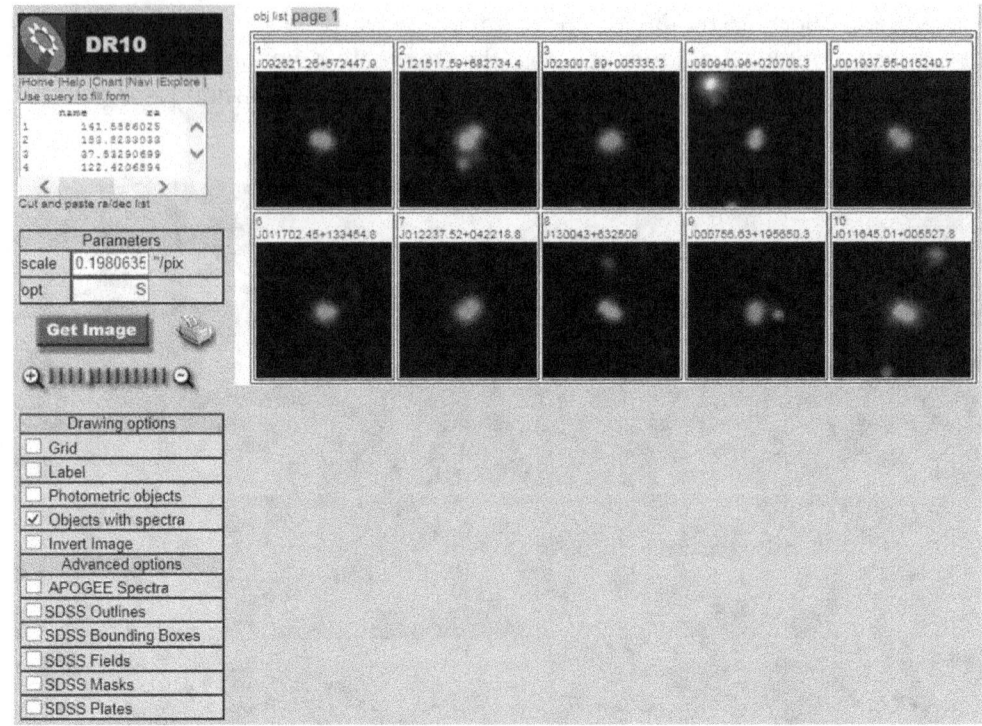

Fig.30 – The Image List page showing the wide range of options that can be selected.

If you are an astronomer hoping to make a discovery the Vizier web site will be a place that you will want to visit very frequently. It is crucial that before announcing any new discovery the researcher has carried out rigorous checks to confirm that the object has not already been listed.

http://vizier.u-strasbg.fr/viz-bin/VizieR

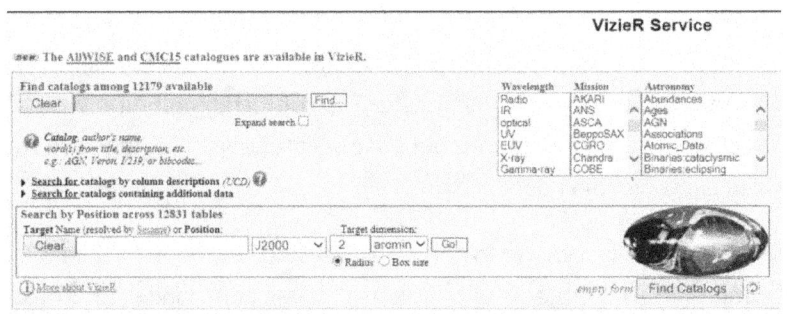

Fig.31 – The Vizier front page

It is equally easy to look for catalogs that cover a specific region of the sky as it is to look for a named catalog. For anybody researching binary or double stars the best place to look is the Washington Visual Double Star Catalog (http://vizier.hia.nrc.ca/viz-bin/VizieR-3?-source=B/wds/).

Results – I have included images plus the key astrometric and photometric information for a sample of 125 "galaxies" that are in reality close pairs of red dwarf stars. These are figures 32 to 43 inclusive.

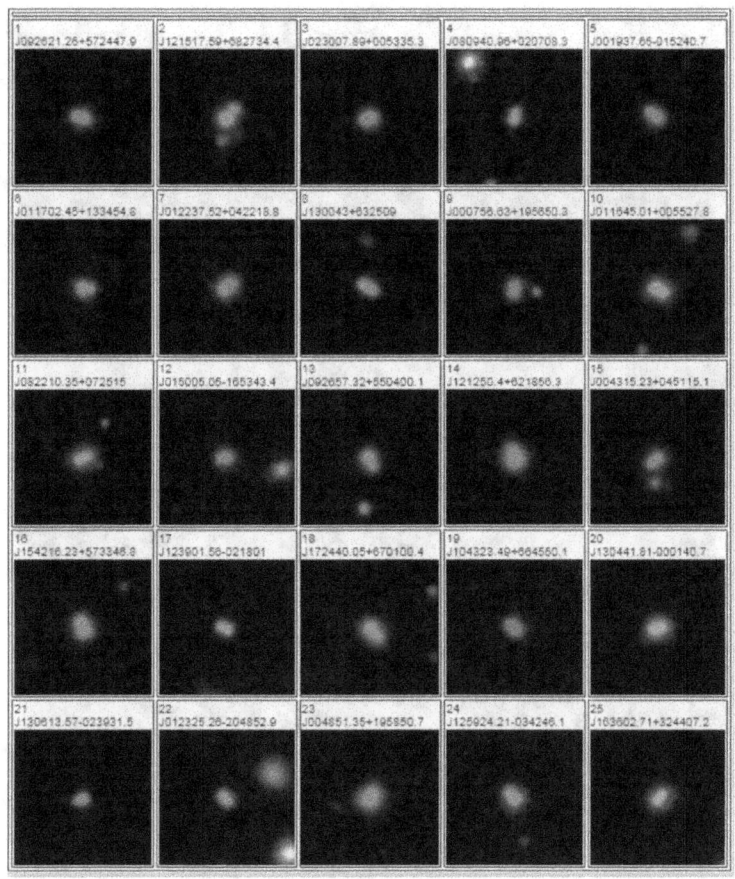

| # | RA | DECL | DEREDDENED PSF MAGNITUDES | | | | | PR in RA | PM in DEC | TOTAL PM |
|---|----|------|---|---|---|---|---|----------|-----------|----------|
| | | | U | G | R | I | Z | MAS/YR | MAS/YR | MAS/YR |
| 1 | 141.589 | 57.413 | 22.185 | 19.662 | 18.219 | 16.781 | 16.039 | -151.09 | -127.08 | 197.43 |
| 2 | 183.823 | 68.460 | 21.530 | 18.910 | 17.590 | 16.469 | 15.886 | 32.74 | -94.67 | 100.17 |
| 3 | 37.533 | 0.893 | 22.770 | 19.491 | 18.170 | 16.598 | 15.764 | -28.27 | -93.11 | 97.31 |
| 4 | 122.421 | 2.119 | 24.062 | 20.338 | 18.942 | 17.933 | 17.376 | -91.90 | -5.77 | 92.08 |
| 5 | 4.907 | -1.878 | 22.585 | 19.976 | 18.591 | 17.205 | 16.444 | 90.30 | -15.25 | 91.58 |
| 6 | 19.260 | 13.582 | 22.113 | 19.251 | 17.752 | 16.531 | 15.836 | -83.35 | -34.07 | 90.04 |
| 7 | 20.656 | 4.372 | 21.955 | 19.409 | 18.023 | 16.641 | 15.907 | 63.57 | -54.10 | 83.48 |
| 8 | 195.179 | 63.419 | 22.375 | 19.986 | 18.602 | 17.387 | 16.801 | -50.62 | -63.64 | 81.32 |
| 9 | 1.986 | 19.947 | 25.518 | 20.445 | 18.963 | 17.359 | 16.421 | 76.59 | 17.57 | 78.58 |
| 10 | 19.188 | 0.924 | 21.123 | 18.439 | 16.980 | 16.125 | 15.637 | 77.81 | -3.57 | 77.89 |
| 11 | 125.543 | 7.421 | 21.535 | 18.853 | 17.552 | 16.498 | 15.692 | -70.14 | 4.20 | 70.27 |
| 12 | 27.521 | -16.895 | 22.420 | 19.924 | 18.538 | 17.060 | 16.234 | 69.70 | 2.94 | 69.76 |
| 13 | 141.739 | 55.067 | 21.862 | 19.711 | 18.341 | 17.065 | 16.392 | -44.22 | -53.80 | 69.64 |
| 14 | 183.210 | 62.316 | 21.487 | 18.971 | 17.575 | 16.359 | 15.659 | 53.29 | -43.83 | 69.00 |
| 15 | 10.813 | 4.854 | 21.765 | 19.523 | 18.181 | 16.711 | 15.871 | 33.57 | 58.89 | 67.79 |
| 16 | 235.568 | 57.563 | 22.803 | 19.620 | 18.092 | 16.784 | 16.163 | -64.39 | -17.13 | 66.63 |
| 17 | 189.757 | -2.300 | 22.448 | 20.206 | 18.969 | 17.879 | 17.302 | -63.11 | -2.82 | 63.17 |
| 18 | 261.167 | 67.017 | 21.715 | 19.279 | 17.937 | 16.839 | 16.310 | -50.59 | -37.16 | 62.77 |
| 19 | 160.848 | 66.764 | 23.933 | 20.453 | 19.097 | 17.436 | 16.462 | -51.64 | -34.87 | 62.30 |
| 20 | 196.174 | -0.028 | 20.818 | 18.556 | 17.273 | 16.393 | 15.911 | -61.84 | 1.35 | 61.85 |
| 21 | 196.557 | -2.659 | 22.916 | 20.734 | 19.258 | 17.922 | 17.102 | -60.94 | 5.42 | 61.18 |
| 22 | 20.855 | -20.815 | 22.527 | 20.513 | 19.072 | 17.811 | 17.256 | 16.63 | -57.97 | 60.30 |
| 23 | 12.214 | 19.981 | 21.395 | 18.619 | 17.193 | 16.095 | 15.492 | -24.08 | -54.40 | 59.49 |
| 24 | 194.851 | -3.713 | 21.658 | 19.132 | 17.775 | 16.613 | 15.900 | -59.47 | -1.14 | 59.48 |
| 25 | 249.011 | 32.735 | 21.196 | 18.539 | 17.165 | 16.334 | 15.793 | -41.82 | 40.68 | 58.35 |

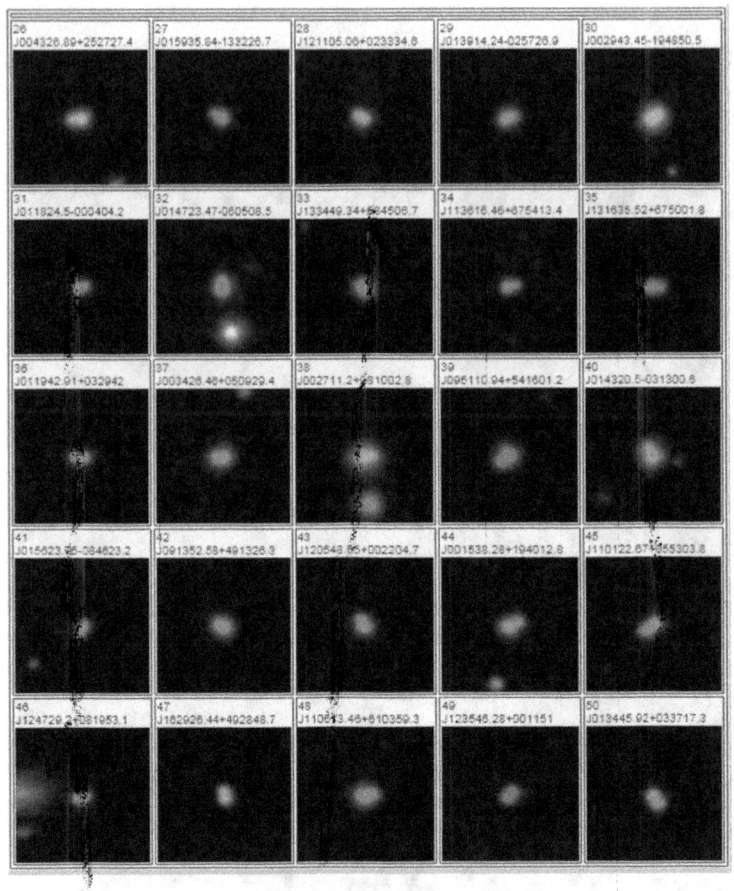

| | | | DEREDDENED PSF MAGNITUDES | | | | | PR in RA | PM in DEC | TOTAL PM |
|---|---|---|---|---|---|---|---|---|---|---|
| # | RA | DECL | U | G | R | I | Z | MAS/YR | MAS/YR | MAS/YR |
| 26 | 10.862 | 25.458 | 21.452 | 19.083 | 17.793 | 17.212 | 16.881 | -29.57 | -47.56 | 56.00 |
| 27 | 29.899 | -13.541 | 22.886 | 20.334 | 18.857 | 17.388 | 16.582 | -7.43 | -55.44 | 55.94 |
| 28 | 182.771 | 2.560 | 22.263 | 19.378 | 18.119 | 17.216 | 16.732 | -12.85 | -54.08 | 55.58 |
| 29 | 24.809 | -2.957 | 21.740 | 19.866 | 18.593 | 17.538 | 16.924 | 2.25 | -53.94 | 53.98 |
| 30 | 7.431 | -19.814 | 20.598 | 18.223 | 16.914 | 16.210 | 15.944 | -33.30 | -42.34 | 53.87 |
| 31 | 19.602 | -0.068 | 23.035 | 20.150 | 18.982 | 17.513 | 16.688 | 19.52 | -50.04 | 53.71 |
| 32 | 26.848 | -6.086 | 21.552 | 19.210 | 17.958 | 16.501 | 15.768 | 52.66 | 7.57 | 53.20 |
| 33 | 203.706 | 58.752 | 21.926 | 19.609 | 18.140 | 16.609 | 15.831 | -28.08 | -45.07 | 53.10 |
| 34 | 174.069 | 67.904 | 22.254 | 20.514 | 19.120 | 17.629 | 16.848 | -48.21 | -20.38 | 52.34 |
| 35 | 199.148 | 67.834 | 22.239 | 20.656 | 19.330 | 17.764 | 16.885 | -19.15 | -48.54 | 52.18 |
| 36 | 19.929 | 3.495 | 22.503 | 19.917 | 18.468 | 17.000 | 16.190 | 34.19 | 39.15 | 51.98 |
| 37 | 8.610 | 5.158 | 21.528 | 18.960 | 17.602 | 16.474 | 15.929 | 51.78 | 0.82 | 51.78 |
| 38 | 6.797 | 8.167 | 20.474 | 18.082 | 16.757 | 15.964 | 15.588 | 23.03 | -46.18 | 51.60 |
| 39 | 147.796 | 54.267 | 21.635 | 19.070 | 17.627 | 16.198 | 15.418 | -45.54 | -21.49 | 50.36 |
| 40 | 25.835 | -3.217 | 21.367 | 18.754 | 17.441 | 16.436 | 15.920 | 49.00 | 9.80 | 49.97 |
| 41 | 29.099 | -8.773 | 22.904 | 20.118 | 18.740 | 17.632 | 16.932 | -16.76 | -46.22 | 49.17 |
| 42 | 138.469 | 49.224 | 23.572 | 19.794 | 18.237 | 16.881 | 16.136 | -40.27 | -28.02 | 49.06 |
| 43 | 181.453 | 0.368 | 21.664 | 19.346 | 17.984 | 16.726 | 16.097 | 25.75 | -41.42 | 48.78 |
| 44 | 3.910 | 19.670 | 22.609 | 19.510 | 18.111 | 17.017 | 16.432 | -6.89 | -47.69 | 48.18 |
| 45 | 165.344 | 5.884 | 22.971 | 20.125 | 18.699 | 17.176 | 16.427 | -7.07 | -47.36 | 47.88 |
| 46 | 191.872 | 8.331 | 22.764 | 20.152 | 18.830 | 17.424 | 16.752 | -33.25 | -34.18 | 47.69 |
| 47 | 247.360 | 49.480 | 22.395 | 19.960 | 18.619 | 17.737 | 17.234 | -25.78 | -39.61 | 47.26 |
| 48 | 166.556 | 61.066 | 21.218 | 18.944 | 17.621 | 16.259 | 15.595 | 4.72 | -46.68 | 46.92 |
| 49 | 188.943 | 0.198 | 23.114 | 20.690 | 19.124 | 17.464 | 16.467 | -45.93 | 9.16 | 46.84 |
| 50 | 23.691 | 3.621 | 23.801 | 20.208 | 18.778 | 17.486 | 16.740 | 33.43 | 32.23 | 46.44 |

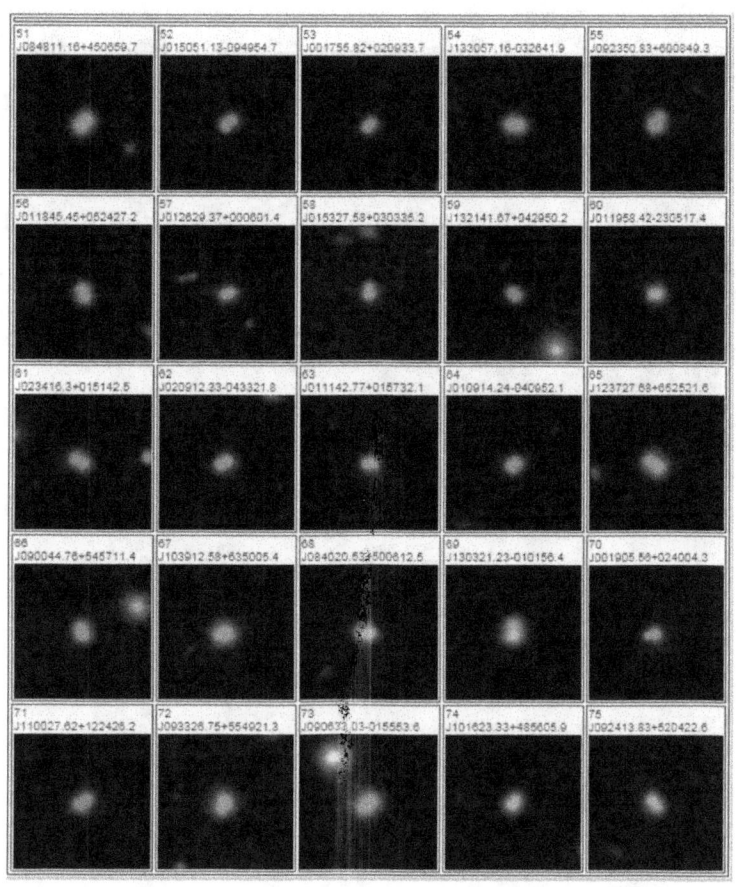

| | | | DEREDDENED PSF MAGNITUDES | | | | | PR in RA | PM in DEC | TOTAL PM |
|---|---|---|---|---|---|---|---|---|---|---|
| # | RA | DECL | U | G | R | I | Z | MAS/YR | MAS/YR | MAS/YR |
| 51 | 132.047 | 45.117 | 23.885 | 19.764 | 18.241 | 17.060 | 16.387 | -44.45 | 1.76 | 44.49 |
| 52 | 27.713 | -9.832 | 22.684 | 20.596 | 19.211 | 17.811 | 16.868 | 43.84 | -0.96 | 43.85 |
| 53 | 4.483 | 2.159 | 23.167 | 20.530 | 19.143 | 17.716 | 16.885 | 43.31 | -3.29 | 43.44 |
| 54 | 202.738 | -3.445 | 22.150 | 19.800 | 18.524 | 17.191 | 16.441 | -43.01 | -0.79 | 43.02 |
| 55 | 140.962 | 60.147 | 22.669 | 19.955 | 18.596 | 17.215 | 16.492 | 2.67 | -42.78 | 42.87 |
| 56 | 19.689 | 5.408 | 24.138 | 20.549 | 18.978 | 17.825 | 17.245 | 37.24 | 20.09 | 42.32 |
| 57 | 21.622 | 0.100 | 23.024 | 20.000 | 18.682 | 17.290 | 16.381 | 34.48 | -24.20 | 42.13 |
| 58 | 28.365 | 3.060 | 22.458 | 20.190 | 18.809 | 17.661 | 17.031 | 41.25 | -4.61 | 41.50 |
| 59 | 200.424 | 4.497 | 23.433 | 20.674 | 19.299 | 17.812 | 16.913 | -36.91 | 18.47 | 41.27 |
| 60 | 19.993 | -23.088 | 24.081 | 19.956 | 18.624 | 17.731 | 17.221 | 20.70 | -35.55 | 41.13 |
| 61 | 38.568 | 1.862 | 22.941 | 20.098 | 18.773 | 17.428 | 16.602 | -16.33 | -37.70 | 41.09 |
| 62 | 32.301 | -4.556 | 22.632 | 19.576 | 18.139 | 17.109 | 16.387 | -11.82 | 39.10 | 40.85 |
| 63 | 17.928 | 1.959 | 22.487 | 20.224 | 18.960 | 17.760 | 17.028 | 36.02 | -17.74 | 40.16 |
| 64 | 17.309 | -4.164 | 22.398 | 19.704 | 18.398 | 17.022 | 16.339 | 34.41 | -20.35 | 39.98 |
| 65 | 189.365 | 65.423 | 21.163 | 18.653 | 17.248 | 16.257 | 15.590 | -39.51 | -0.33 | 39.51 |
| 66 | 135.187 | 54.953 | 21.798 | 19.260 | 17.893 | 16.659 | 15.897 | -33.98 | -20.05 | 39.46 |
| 67 | 159.802 | 63.835 | 21.344 | 18.716 | 17.301 | 16.189 | 15.595 | 36.23 | -14.98 | 39.20 |
| 68 | 130.086 | 50.103 | 22.335 | 19.766 | 18.170 | 17.345 | 16.870 | -22.20 | -32.26 | 39.16 |
| 69 | 195.838 | -1.032 | 21.409 | 19.077 | 17.792 | 17.093 | 16.725 | -38.63 | -6.29 | 39.13 |
| 70 | 4.773 | 2.668 | 21.997 | 19.917 | 18.548 | 17.294 | 16.602 | -19.82 | -33.60 | 39.01 |
| 71 | 165.115 | 12.407 | 22.135 | 19.960 | 18.750 | 17.311 | 16.463 | -11.14 | -37.36 | 38.99 |
| 72 | 143.361 | 55.823 | 21.276 | 18.738 | 17.433 | 16.429 | 15.891 | 22.82 | -30.84 | 38.36 |
| 73 | 136.638 | -1.932 | 22.062 | 19.049 | 17.652 | 16.411 | 15.773 | 3.02 | -38.11 | 38.23 |
| 74 | 154.097 | 48.935 | 22.089 | 19.482 | 18.138 | 17.342 | 16.843 | 2.56 | -37.72 | 37.81 |
| 75 | 141.058 | 52.073 | 21.976 | 19.457 | 18.223 | 17.316 | 16.768 | 2.15 | -37.74 | 37.80 |

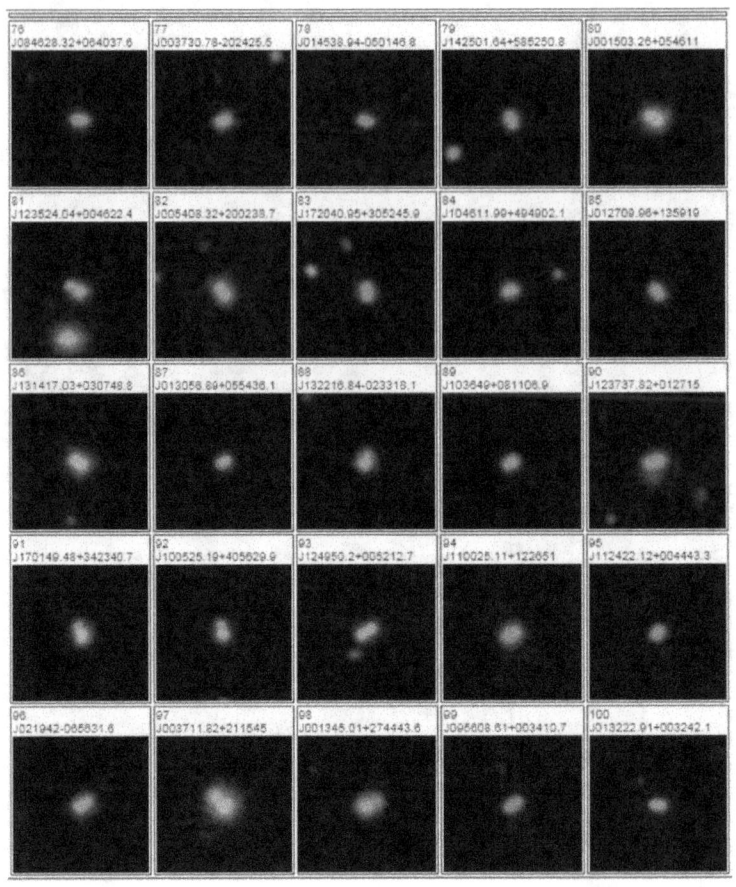

| # | RA | DECL | DEREDDENED PSF MAGNITUDES | | | | | PR in RA | PM in DEC | TOTAL PM |
|---|-----|------|-----|-----|-----|-----|-----|----------|-----------|----------|
| | | | U | G | R | I | Z | MAS/YR | MAS/YR | MAS/YR |
| 76 | 131.618 | 6.677 | 22.917 | 20.119 | 18.777 | 17.445 | 16.702 | -30.16 | -22.56 | 37.66 |
| 77 | 9.378 | -20.407 | 23.794 | 20.143 | 18.759 | 17.591 | 16.947 | -5.34 | -36.76 | 37.14 |
| 78 | 26.412 | -5.030 | 23.104 | 20.480 | 19.172 | 17.722 | 16.877 | 30.73 | -20.27 | 36.81 |
| 79 | 216.257 | 58.881 | 23.519 | 20.389 | 19.309 | 17.757 | 17.000 | -32.89 | 16.51 | 36.80 |
| 80 | 3.764 | 5.770 | 20.705 | 18.329 | 16.993 | 16.059 | 15.537 | 34.06 | -13.72 | 36.72 |
| 81 | 188.850 | 0.773 | 22.175 | 19.613 | 18.175 | 17.125 | 16.529 | -34.30 | -9.94 | 35.71 |
| 82 | 13.535 | 20.044 | 21.839 | 19.337 | 17.930 | 16.873 | 16.310 | -21.42 | -28.31 | 35.51 |
| 83 | 260.171 | 30.879 | 22.038 | 19.840 | 18.616 | 17.268 | 16.533 | -33.51 | -10.88 | 35.23 |
| 84 | 161.550 | 49.817 | 23.381 | 20.355 | 18.801 | 17.544 | 16.810 | -5.87 | -34.61 | 35.11 |
| 85 | 21.792 | 13.989 | 22.469 | 20.357 | 19.025 | 17.747 | 17.059 | -0.97 | -35.06 | 35.07 |
| 86 | 198.571 | 3.130 | 21.186 | 18.695 | 17.331 | 16.597 | 16.195 | -19.89 | -28.78 | 34.98 |
| 87 | 22.737 | 5.910 | 22.317 | 20.301 | 19.029 | 17.659 | 16.784 | 24.43 | -25.01 | 34.96 |
| 88 | 200.570 | -2.555 | 21.972 | 19.623 | 18.366 | 17.255 | 16.672 | -34.87 | 0.37 | 34.87 |
| 89 | 159.204 | 8.185 | 22.416 | 20.384 | 19.045 | 17.891 | 17.203 | -10.92 | -33.07 | 34.82 |
| 90 | 189.408 | 1.454 | 22.656 | 19.612 | 18.233 | 16.851 | 16.040 | -32.55 | 12.13 | 34.74 |
| 91 | 255.456 | 34.395 | 21.741 | 19.283 | 17.883 | 17.141 | 16.765 | -17.37 | -29.07 | 33.86 |
| 92 | 151.355 | 40.942 | 21.901 | 19.634 | 18.304 | 17.437 | 16.950 | -16.94 | -29.31 | 33.85 |
| 93 | 192.459 | 0.870 | 22.880 | 19.811 | 18.403 | 17.450 | 16.852 | 13.53 | -30.95 | 33.77 |
| 94 | 165.105 | 12.448 | 22.120 | 19.625 | 18.251 | 16.885 | 16.105 | -33.33 | 4.88 | 33.69 |
| 95 | 171.092 | 0.745 | 22.612 | 20.468 | 19.078 | 17.612 | 16.841 | -32.63 | -5.90 | 33.16 |
| 96 | 34.925 | -6.942 | 21.686 | 19.185 | 17.847 | 16.695 | 16.090 | 26.49 | -18.40 | 32.25 |
| 97 | 9.299 | 21.263 | 20.562 | 18.091 | 16.693 | 15.943 | 15.454 | 29.61 | -11.96 | 31.93 |
| 98 | 3.438 | 27.745 | 23.472 | 19.596 | 18.249 | 17.157 | 16.470 | 31.54 | -1.60 | 31.58 |
| 99 | 149.036 | 0.570 | 22.616 | 20.542 | 19.040 | 17.600 | 16.823 | -28.92 | -12.36 | 31.45 |
| 100 | 23.095 | 0.545 | 22.389 | 20.289 | 18.974 | 17.707 | 16.916 | -10.10 | -29.76 | 31.43 |

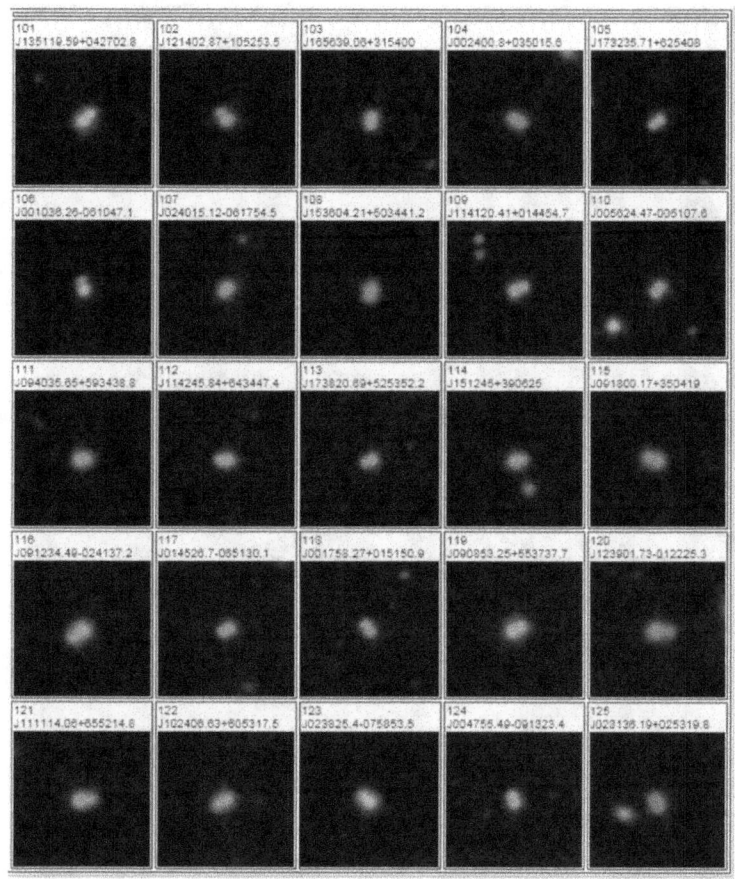

| | | | DEREDDENED PSF MAGNITUDES | | | | | PR in RA | PM in DEC | TOTAL PM |
|---|---|---|---|---|---|---|---|---|---|---|
| # | RA | DECL | U | G | R | I | Z | MAS/YR | MAS/YR | MAS/YR |
| 101 | 207.832 | 4.451 | 24.919 | 19.145 | 17.763 | 16.938 | 16.494 | 23.60 | -20.73 | 31.41 |
| 102 | 183.512 | 10.882 | 22.061 | 20.127 | 18.681 | 17.452 | 16.776 | 7.28 | -30.33 | 31.19 |
| 103 | 254.163 | 31.900 | 22.077 | 20.155 | 18.817 | 17.766 | 17.221 | -2.15 | -30.91 | 30.98 |
| 104 | 6.003 | 3.838 | 22.907 | 20.286 | 18.807 | 17.377 | 16.594 | 30.60 | -1.61 | 30.64 |
| 105 | 263.149 | 62.902 | 22.558 | 20.215 | 18.859 | 17.719 | 17.082 | -23.74 | 19.37 | 30.64 |
| 106 | 2.651 | -6.180 | 21.715 | 19.802 | 18.474 | 17.712 | 17.280 | 17.80 | -24.92 | 30.62 |
| 107 | 40.063 | -6.298 | 22.183 | 19.882 | 18.635 | 17.305 | 16.560 | 27.66 | 12.54 | 30.37 |
| 108 | 234.018 | 50.578 | 23.218 | 20.362 | 19.076 | 17.684 | 17.155 | 1.90 | 30.28 | 30.34 |
| 109 | 175.335 | 1.749 | 22.592 | 19.512 | 18.296 | 17.254 | 16.610 | 22.81 | -19.90 | 30.27 |
| 110 | 14.102 | -0.852 | 22.003 | 19.737 | 18.361 | 17.255 | 16.471 | -15.05 | -26.22 | 30.23 |
| 111 | 145.149 | 59.577 | 22.185 | 19.656 | 18.363 | 17.175 | 16.479 | 13.66 | -26.57 | 29.88 |
| 112 | 175.691 | 64.580 | 22.226 | 19.405 | 18.011 | 16.992 | 16.423 | -28.55 | 8.51 | 29.79 |
| 113 | 264.586 | 52.898 | 22.392 | 20.261 | 18.931 | 17.891 | 17.281 | -23.46 | 18.05 | 29.60 |
| 114 | 228.188 | 39.107 | 22.461 | 20.014 | 18.716 | 17.613 | 16.971 | -2.20 | -28.96 | 29.04 |
| 115 | 139.501 | 35.072 | 23.367 | 20.358 | 18.939 | 17.493 | 16.800 | -25.11 | -13.97 | 28.74 |
| 116 | 138.144 | -2.694 | 22.395 | 19.584 | 18.237 | 17.135 | 16.537 | -28.44 | 1.84 | 28.50 |
| 117 | 26.361 | -6.858 | 25.344 | 20.142 | 18.795 | 17.612 | 16.769 | -2.59 | -28.36 | 28.48 |
| 118 | 4.493 | 1.864 | 22.380 | 20.505 | 19.240 | 17.929 | 17.184 | 25.04 | -12.80 | 28.12 |
| 119 | 137.222 | 55.627 | 21.153 | 18.673 | 17.253 | 16.523 | 16.109 | 27.83 | 3.04 | 28.00 |
| 120 | 189.757 | -1.374 | 22.981 | 20.363 | 18.779 | 17.087 | 16.254 | -22.07 | -17.00 | 27.86 |
| 121 | 167.809 | 65.871 | 22.474 | 19.684 | 18.257 | 17.173 | 16.578 | 23.84 | -8.65 | 25.36 |
| 122 | 156.028 | 60.888 | 22.268 | 20.196 | 18.866 | 17.621 | 17.049 | -20.90 | -13.71 | 25.00 |
| 123 | 39.606 | -7.982 | 21.230 | 19.102 | 17.945 | 17.266 | 16.924 | 11.42 | -21.92 | 24.72 |
| 124 | 11.981 | -9.223 | 22.527 | 19.929 | 18.613 | 17.754 | 17.314 | 2.98 | -24.52 | 24.70 |
| 125 | 37.901 | 2.889 | 22.568 | 20.367 | 18.854 | 17.030 | 16.087 | -23.12 | -8.55 | 24.65 |

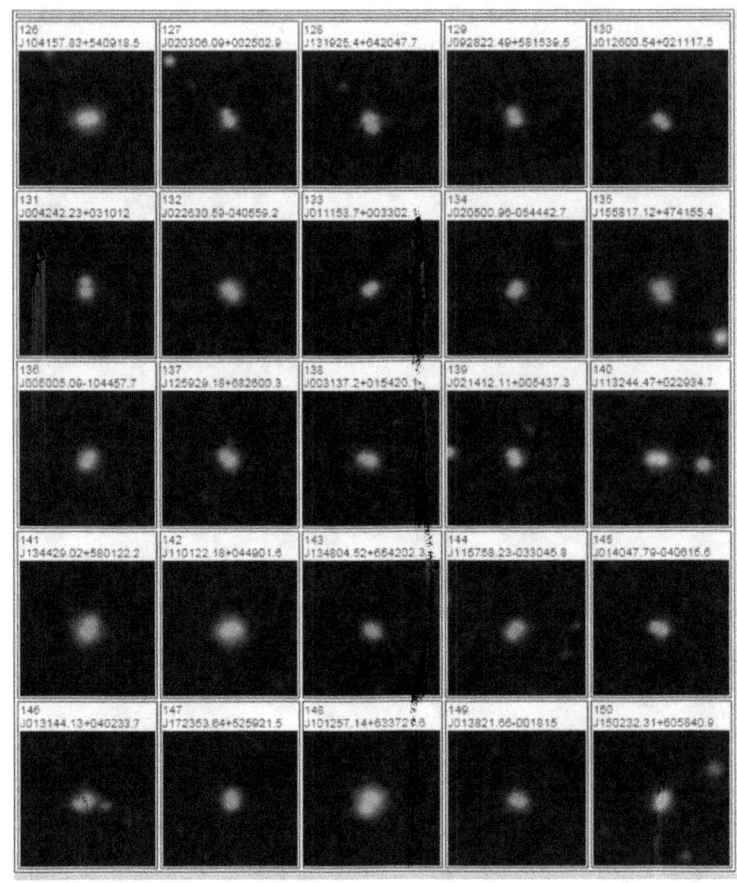

| | | | DEREDDENED PSF MAGNITUDES | | | | | PR in RA | PM in DEC | TOTAL PM |
|---|---|---|---|---|---|---|---|---|---|---|
| # | RA | DECL | U | G | R | I | Z | MAS/YR | MAS/YR | MAS/YR |
| 126 | 160.491 | 54.155 | 21.130 | 18.529 | 17.174 | 16.419 | 15.975 | -18.27 | -16.54 | 24.64 |
| 127 | 30.775 | 0.417 | 21.625 | 19.560 | 18.174 | 17.119 | 16.598 | -6.10 | -23.80 | 24.57 |
| 128 | 199.856 | 64.347 | 22.696 | 19.999 | 18.656 | 17.473 | 16.855 | -18.03 | 16.69 | 24.57 |
| 129 | 142.094 | 58.261 | 23.091 | 20.294 | 18.883 | 17.825 | 17.178 | 18.35 | -16.26 | 24.51 |
| 130 | 21.502 | 2.188 | 22.574 | 20.269 | 18.894 | 17.760 | 17.171 | 23.73 | -5.25 | 24.30 |
| 131 | 10.676 | 3.170 | 23.452 | 20.311 | 18.879 | 17.675 | 16.991 | 11.68 | -20.83 | 23.88 |
| 132 | 36.627 | -4.100 | 22.213 | 19.668 | 18.137 | 17.086 | 16.437 | 23.28 | -3.72 | 23.58 |
| 133 | 17.974 | 0.551 | 22.644 | 20.320 | 18.959 | 17.866 | 17.305 | -1.99 | -23.49 | 23.57 |
| 134 | 31.254 | -5.745 | 22.875 | 20.350 | 19.044 | 17.731 | 17.035 | -14.58 | -18.52 | 23.57 |
| 135 | 239.571 | 47.699 | 21.907 | 19.707 | 18.310 | 17.002 | 16.317 | -22.70 | 6.34 | 23.57 |
| 136 | 12.521 | -10.749 | 22.024 | 19.719 | 18.366 | 17.388 | 16.786 | 20.13 | -10.75 | 22.82 |
| 137 | 194.872 | 68.433 | 22.462 | 20.074 | 18.602 | 17.558 | 16.910 | -22.20 | -4.42 | 22.64 |
| 138 | 7.905 | 1.906 | 22.732 | 20.081 | 18.654 | 17.442 | 16.765 | -3.65 | -22.08 | 22.38 |
| 139 | 33.550 | 0.910 | 23.320 | 20.008 | 18.565 | 17.646 | 17.145 | 13.46 | -17.40 | 22.00 |
| 140 | 173.185 | 2.493 | 22.187 | 19.445 | 18.124 | 17.512 | 17.137 | -20.05 | 8.90 | 21.93 |
| 141 | 206.121 | 58.023 | 21.396 | 18.689 | 17.283 | 15.987 | 15.313 | 20.27 | -8.22 | 21.88 |
| 142 | 165.342 | 4.817 | 20.972 | 18.324 | 17.055 | 16.052 | 15.451 | -20.09 | 8.63 | 21.87 |
| 143 | 207.019 | 65.701 | 22.105 | 20.535 | 19.225 | 17.793 | 17.046 | 10.61 | -18.99 | 21.75 |
| 144 | 179.493 | -3.513 | 22.909 | 20.110 | 18.798 | 17.387 | 16.637 | -14.64 | -15.98 | 21.68 |
| 145 | 25.199 | -4.104 | 22.213 | 19.985 | 18.675 | 17.513 | 16.965 | -10.12 | -19.10 | 21.62 |
| 146 | 22.934 | 4.043 | 22.126 | 20.090 | 18.696 | 17.471 | 16.767 | -19.96 | -8.28 | 21.61 |
| 147 | 260.974 | 52.989 | 22.149 | 20.244 | 18.887 | 17.565 | 16.816 | -11.33 | -18.34 | 21.56 |
| 148 | 153.238 | 63.623 | 21.068 | 18.603 | 17.160 | 16.277 | 15.758 | 21.25 | 3.28 | 21.50 |
| 149 | 24.590 | -0.304 | 21.922 | 19.371 | 18.073 | 17.047 | 16.312 | 0.22 | -21.48 | 21.48 |
| 150 | 225.635 | 60.978 | 22.632 | 19.431 | 18.113 | 17.226 | 16.697 | 7.44 | -19.95 | 21.30 |

Two features of these results are noteworthy. These false galaxy pairs are not resolved in the widely used Digitized Sky Survey plates and so they don't draw themselves to the attention of any astronomer hunting for uncatalogued systems.

Fig. 44 - False galaxy #16 – unresolved in this 5 x 5 arc minute
image from the Digitized Sky Survey

These pairs can be found across a wide range of proper motions. A total proper motion of 60mas/yr has long been suggested as the minimum value that can be used as a diagnostic tool yet there are many examples of close pairs where the total proper motion is far less than this.

Exercise 2 – A statistical approach to double star discovery

The SQL programme shown below selects all the stars (type = 6) that are primary objects (mode = 1) and that have clean photometry (clean = 1) and that are in a chosen area of the sky. Only objects with the photometric characteristics of red dwarf stars form part of the output file.

```
SELECT TOP 1000000  s1.ra as RA1, s1.dec as DE1, p1.pmra as PM1a, p1.pmdec as PM1b,
s1.psfMag_u as U, s1.psfMag_g as G, s1.psfMag_r as R, s1.psfMag_i as I, s1.psfMag_z as Z,
s1.extinction_u as UE, s1.extinction_g as GE, s1.extinction_r as RE, s1.extinction_i as IE,
s1.extinction_z as ZE

INTO mydb.SummaryB4NOPM
FROM PhotoObjAll S1, ProperMotions p1
WHERE S1.ObjID = p1.ObjID

AND S1.ra between 170 and 190
AND S1.dec between 0 and 20
AND S1.Mode = 1
AND S1.Type = 6
AND S1.Clean = 1

AND s1.dered_r between 14.1 and 22.3
AND s1.dered_i between 13.8 and 21.7
AND s1.dered_z between 12.3 and 20.1
AND (s1.dered_r-s1.dered_i) between 0.62 and 2.82
AND (s1.dered_i-s1.dered_z) between 0.32 and 1.85
```

A total of 14,365 double stars were identified with a separation between the two components of between 1 and 20 arc seconds.

It is crucial when doing the analysis to allow for the fact that the area of sky between 1 and 2 arc seconds from a star is <u>not</u> the same as the area between 2 and 3 arc seconds and so on. When the area correction factor is applied to the results some interesting trends emerge.

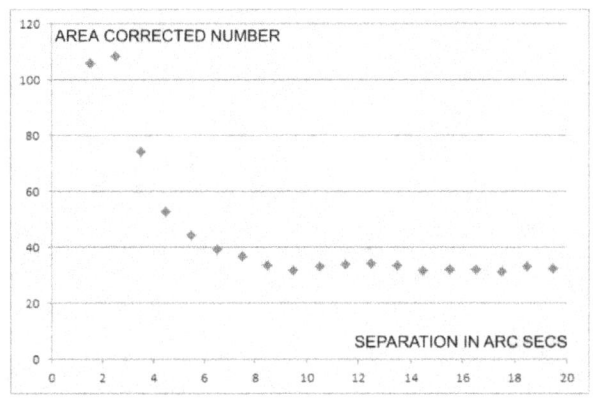

Fig. 45 – The distribution of double stars by separation

Once the separation reaches 8 arc seconds the number of double star systems detected per unit area of sky remains almost constant. It is possible that this "background level" represents random "line of sight" systems and that everything above this level is due to genuine binary systems where the two stars are gravitationally linked.

| RADIUS | | AREA OF | | | |
|---|---|---|---|---|---|
| INNER | OUTER | TORUS | RELATIVE | NUMBER OF | AREA CORRECTED |
| ARC SEC | ARC SEC | SQR ARC SEC | AREA | DOUBLE STARS | NUMBER |
| 0 | 1 | 3.1416 | 1 | | |
| 1 | 2 | 9.4248 | 3 | 318 | 106.0 |
| 2 | 3 | 15.708 | 5 | 543 | 108.6 |
| 3 | 4 | 21.9912 | 7 | 519 | 74.1 |
| 4 | 5 | 28.2744 | 9 | 476 | 52.9 |
| 5 | 6 | 34.5576 | 11 | 488 | 44.4 |
| 6 | 7 | 40.8408 | 13 | 512 | 39.4 |
| 7 | 8 | 47.124 | 15 | 554 | 36.9 |
| 8 | 9 | 53.4072 | 17 | 571 | 33.6 |
| 9 | 10 | 59.6904 | 19 | 600 | 31.6 |
| 10 | 11 | 65.9736 | 21 | 699 | 33.3 |
| 11 | 12 | 72.2568 | 23 | 777 | 33.8 |
| 12 | 13 | 78.54 | 25 | 857 | 34.3 |
| 13 | 14 | 84.8232 | 27 | 906 | 33.6 |
| 14 | 15 | 91.1064 | 29 | 916 | 31.6 |
| 15 | 16 | 97.3896 | 31 | 992 | 32.0 |
| 16 | 17 | 103.6728 | 33 | 1063 | 32.2 |
| 17 | 18 | 109.956 | 35 | 1090 | 31.1 |
| 18 | 19 | 116.2392 | 37 | 1220 | 33.0 |
| 19 | 20 | 122.5224 | 39 | 1264 | 32.4 |

Fig. 46 – The area corrected distribution of double stars

For separations from 1 to 3 arc seconds the density of double stars is three times the background level - which suggests that 2/3rds of the systems at these separations are genuine.

Based on the work by West et al (2011) it was possible to assign spectral classes to the 315,000 + individual stars whose resulted were downloaded.

| Spectral Type M+ | SDSS (r-i) | SDSS (i-z) | SDSS/2MASS (z-J) | 2MASS (H-K) | | | SDSS (r-i) | SDSS (i-z) | SDSS/2MASS (z-J) | 2MASS (H-K) |
|---|---|---|---|---|---|---|---|---|---|---|
| 0 | 0.56 | 0.33 | 1.20 | 0.16 | Range | | 1.14 | 1.38 | 0.95 | 0.26 |
| 1 | 0.73 | 0.41 | 1.32 | 0.19 | Sensitivity | | 4.38 | 5.31 | 3.65 | 1.00 |
| 2 | 0.96 | 0.53 | 1.26 | 0.22 | | | | | | |
| 3 | 1.13 | 0.61 | 1.30 | 0.23 | | | | | | |
| 4 | 1.33 | 0.71 | 1.37 | 0.25 | Median colours (West el al, 2011) | | | | | |
| 5 | 1.62 | 0.90 | 1.47 | 0.28 | | | | | | |
| 6 | 1.92 | 1.05 | 1.61 | 0.31 | | | | | | |
| 7 | 2.09 | 1.14 | 1.71 | 0.34 | | | | | | |
| 8 | 2.56 | 1.41 | 1.93 | 0.39 | | | | | | |
| 9 | 2.70 | 1.71 | 2.15 | 0.42 | | | | | | |

Fig. 47 – Spectral classification by SDSS colour

The Sloan Digital Sky Survey Data Release 7 Spectroscopic M Dwarf Catalog I: Data

http://arxiv.org/abs/1101.1082

West et al (2011) - The Astronomical Journal, Volume 141, Issue 3, article id. 97

The distribution between the different M type sub-classes is very uneven. M7 to M9 red dwarf stars are intrinsically so faint that they can only be detected when very nearby (astronomically speaking). Even with a large scope and a sensitive CCD examples are hard to find.

| SPECTRAL TYPE | NUMBER OF STARS | PERCENTAGE OF STARS |
|---|---|---|
| M0 | 8464 | 2.68 |
| M1 | 60849 | 19.29 |
| M2 | 69504 | 22.03 |
| M3 | 78857 | 25.00 |
| M4 | 63753 | 20.21 |
| M5 | 25352 | 8.04 |
| M6 | 6436 | 2.04 |
| M7 | 1734 | 0.55 |
| M8 | 469 | 0.15 |
| M9 | 71 | 0.02 |

Fig. 48 – Spectral classification by number

The paper by Bochanski (2010), plus the subsequent typographical correction, made it possible to determine the absolute magnitude – and hence the distance – of the surveyed red dwarf stars. However the large standard deviation limits the use of distance as a tool to identify stars that are a similar distance from the observer.

Color–Absolute Magnitude Relations in the *ugriz* System

| Absolute Magnitude | Color Range | Best Fit | $\sigma_{M_r}$ |
|---|---|---|---|
| $M_r$ | $0.50 < r - z < 4.53$ | $5.190 + 2.474\,(r - z) + 0.4340\,(r - z)^2 - 0.08635\,(r - z)^3$ | 0.394 |
| $M_r$ | $0.62 < r - i < 2.82$ | $5.025 + 4.548\,(r - i) + 0.4175\,(r - i)^2 - 0.18315\,(r - i)^3$ | 0.403 |
| $M_r$ | $0.32 < i - z < 1.85$ | $4.748 + 8.275\,(i - z) + 2.2789\,(i - z)^2 - 1.5337\,(i - z)^3$ | 0.481 |

Fig. 49 – The colour / absolute magnitude relationship

The Luminosity and Mass Functions of Low-Mass Stars in the Galactic Disk: II. The Field

http://arxiv.org/abs/1004.4002

Bochanski et al (2010) - The Astronomical Journal, Volume 139, Issue 6, article id. 2679-2699

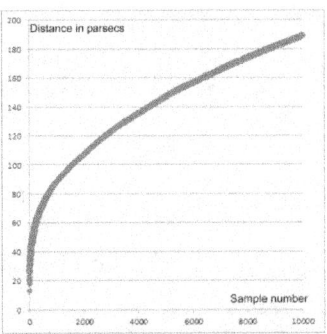

Fig. 50 – The estimated distances of the sampled red dwarf stars

There are few nearby stars in this sample because a nearby red dwarf star with an absolute magnitude of 10 will saturate the SDSS detectors and so would not be classified as having "good photometry".

The next programme shows how useful results can sometimes be obtained serendipitously. The plan was to identify systems with closely matching and high quality proper motion results.

```
SELECT TOP 100000  s1.ra as RA1, s1.dec as DE1, p1.pmra as PM1a, p1.pmdec as PM1b,
p2.pmra as PM2a, p2.pmdec as PM2b, s1.psfMag_u as U, s1.psfMag_g as G, s1.psfMag_r as R,
s1.psfMag_i as I, s1.psfMag_z as Z, s1.extinction_u as UE, s1.extinction_g as GE, s1.extinction_r
as RE, s1.extinction_i as IE, s1.extinction_z as ZE, s2.ra as RA2, s2.dec as DE2, s2.psfMag_u as
U2, s2.psfMag_g as G2, s2.psfMag_r as R2, s2.psfMag_i as I2, s2.psfMag_z as Z2,
s2.extinction_u as UE2, s2.extinction_g as GE2, s2.extinction_r as RE2, s2.extinction_i as IE2,
s2.extinction_z as ZE2, N.distance as DI

INTO mydb.BookPMv1
FROM PhotoObjAll S1, Neighbors N, PhotoObjAll S2, ProperMotions p1, ProperMotions p2
WHERE S1.ObjID = N.ObjID
AND S2.ObjID = N.NeighborObjID
AND s1.ObjID = p1.ObjID
AND s2.ObjID = p2.ObjID

AND S1.ra between 170 and 190
AND S1.dec between 0 and 20

AND S1.Mode = 1
AND S2.Mode = 1
AND S1.Type = 6
AND S2.Type = 6
AND S1.Clean = 1
AND S2.Clean = 1

AND N.DISTANCE between 0.05 and 0.5

AND p1.MATCH = 1
AND p1.NFIT >4
AND p1.DIST22 > 7
AND p1.sigRA < 525
AND p1.sigDEC < 525
AND (p1.pmra*p1.pmra) + (p1.pmdec*p1.pmdec) > 1600

AND p2.MATCH = 1
AND p2.NFIT >4
AND p2.DIST22 > 7
AND p2.sigRA < 525
AND p2.sigDEC < 525
AND (p2.pmra*p2.pmra) + (p2.pmdec*p2.pmdec) > 1600

AND abs(p1.pmra - p2.pmra) < (p1.pmraerr + p2.pmraerr)
AND abs(p1.pmdec - p2.pmdec) < (p1.pmdecerr + p2.pmdecerr)
```

Although the programme succeeded in identifying a number of such systems none were closer together than 8 arc seconds. This suggested that the formal SDSS designation of "high quality" proper motion could not be applied to stars with another star close by which would always be the case with a close double. Removing both the DIST22 > 7 commands reduced the separation of the closest detectable system to 6.7 arc seconds and further removing both the sigRA and both the sigDEC commands reduced the minimum separation down to 3 arc seconds.

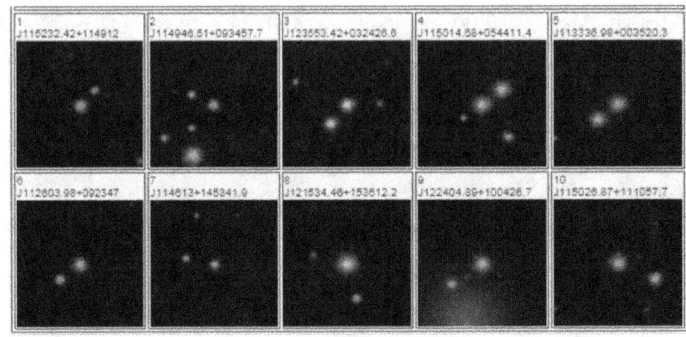

Fig. 51 and Fig. 52 – Red dwarf stars with matching proper motions

| | PRIMARY | | SECONDARY | | PROPER MOTION mas/yr PRIMARY | | PROPER MOTION mas/yr PRIMARY | | SEPARATION |
|---|---|---|---|---|---|---|---|---|---|
| # | RA1 | DE1 | RA2 | DE2 | PM1a | PM1b | PM2a | PM2b | ARC SEC |
| 1 | 178.1351 | 11.8200 | 178.1335 | 11.8216 | 9.3 | -45.0 | 11.0 | -44.6 | 8.02 |
| 2 | 177.4438 | 9.5827 | 177.4462 | 9.5838 | 10.2 | -45.5 | 10.1 | -45.9 | 9.20 |
| 3 | 188.9726 | 3.4074 | 188.9743 | 3.4054 | -48.9 | 14.6 | -46.5 | 14.5 | 9.34 |
| 4 | 177.5612 | 5.7365 | 177.5591 | 5.7381 | 27.7 | -41.2 | 28.7 | -38.5 | 9.35 |
| 5 | 173.4041 | 0.5890 | 173.4062 | 0.5874 | -51.7 | -9.6 | -50.6 | -10.0 | 9.51 |
| 6 | 171.5166 | 9.3964 | 171.5188 | 9.3948 | -48.9 | -4.4 | -45.2 | -2.6 | 9.55 |
| 7 | 176.5542 | 14.8950 | 176.5573 | 14.8957 | -41.1 | -25.5 | -34.5 | -27.7 | 11.13 |
| 8 | 183.8936 | 15.6034 | 183.8926 | 15.5998 | -38.7 | -38.1 | -37.4 | -38.9 | 13.51 |
| 9 | 186.0204 | 10.0741 | 186.0236 | 10.0720 | -43.6 | 11.9 | -37.4 | 14.9 | 13.72 |
| 10 | 177.6120 | 11.1827 | 177.6081 | 11.1811 | -40.9 | -13.5 | -41.1 | -9.3 | 14.84 |

There are a number of possible alternatives to the filter used to define what difference between the proper motion of the two components is "acceptable".

The programme shown above uses:-

AND abs(p1.pmra - p2.pmra) < (p1.pmraerr + p2.pmraerr)
AND abs(p1.pmdec - p2.pmdec) < (p1.pmdecerr + p2.pmdecerr)

However either of the options shown below are possible substitutes.

| AND (p1.pmra / p2.pmra) between 0.9 and 1.1<br>AND (p1.pmdec / p2.pmdec) between 0.9 and 1.1 | AND abs(p1.pmra / p2.pmra) < 5<br>AND abs(p1.pmdec / p2.pmdec) < 5 |
|---|---|

Fig. 53 – Alternative approaches to matching proper motions

Matching close neighbours both by distance and by similarities in their spectral types is another possible way to identify previously unreported double star systems.

As shown previously using the method described by West et al (2011) it was possible to assign spectral classes to individual red dwarf stars.

| Spectral Type M+ | SDSS (r-i) | SDSS (i-z) | SDSS/2MASS (z-J) | 2MASS (H-K) | | SDSS (r-i) | SDSS (i-z) | SDSS/2MASS (z-J) | 2MASS (H-K) |
|---|---|---|---|---|---|---|---|---|---|
| 0 | 0.56 | 0.33 | 1.20 | 0.16 | Range | 1.14 | 1.38 | 0.95 | 0.26 |
| 1 | 0.73 | 0.41 | 1.32 | 0.19 | Sensitivity | 4.38 | 5.31 | 3.65 | 1.00 |
| 2 | 0.96 | 0.53 | 1.26 | 0.22 | | | | | |
| 3 | 1.13 | 0.61 | 1.30 | 0.23 | | | | | |
| 4 | 1.33 | 0.71 | 1.37 | 0.25 | | Median colours (West el al, 2011) | | | |
| 5 | 1.62 | 0.90 | 1.47 | 0.28 | | | | | |
| 6 | 1.92 | 1.05 | 1.61 | 0.31 | | | | | |
| 7 | 2.09 | 1.14 | 1.71 | 0.34 | | | | | |
| 8 | 2.56 | 1.41 | 1.93 | 0.39 | | | | | |
| 9 | 2.70 | 1.71 | 2.15 | 0.42 | | | | | |

### Fig. 54 – Matching the SDSS colour to the spectral type

The Sloan Digital Sky Survey Data Release 7 Spectroscopic M Dwarf Catalog I: Data

http://arxiv.org/abs/1101.1082

West et al (2011) - The Astronomical Journal, Volume 141, Issue 3, article id. 97

```
SELECT TOP 100 s1.ra as RA, s1.dec as DE, s1.psfMag_u as U1, s1.psfMag_g as G1, s1.psfMag_r as R1, s1.psfMag_i as I1,
s1.psfMag_z as Z1, s1.extinction_u as UE1, s1.extinction_g as GE1, s1.extinction_r as RE1, s1.extinction_i as IE1, s1.extinction_z
as ZE1, s2.ra as RA2, s2.dec as DE2, s2.psfMag_u as U2, s2.psfMag_g as G2, s2.psfMag_r as R2, s2.psfMag_i as I2, s2.psfMag_z
as Z2, s2.extinction_u as UE2, s2.extinction_g as GE2, s2.extinction_r as RE2, s2.extinction_i as IE2, s2.extinction_z as ZE2,
N.distance as DIST

INTO mydb.SpectrumTwinsB
FROM PhotoObjAll S1, Neighbors N, PhotoObjAll S2
WHERE S1.ObjID = N.ObjID
AND S2.ObjID = N.NeighborObjID

AND S1.Mode = 1
AND S2.Mode = 1
AND S1.Type = 6
AND S2.Type = 6
AND S1.Clean = 1
AND S2.Clean = 1
AND N.DISTANCE between 0.05 and 0.5
AND (s1.psfMag_r - s1.extinction_r) between 14 and 17
AND (s1.psfMag_i - s1.extinction_i) between 14 and 17
AND (s1.psfMag_z - s1.extinction_z) between 12 and 15
AND (s2.psfMag_r - s2.extinction_r) between 14 and 17
AND (s2.psfMag_i - s2.extinction_i) between 14 and 17
AND (s2.psfMag_z - s2.extinction_z) between 12 and 15

AND -0.606*(s1.dered_r-s1.dered_i)*(s1.dered_r-s1.dered_i) + 6.0251*(s1.dered_r-s1.dered_i) - 3.1214 between -0.5 and 9.5
AND -3.1554*(s1.dered_i-s1.dered_z)*(s1.dered_i-s1.dered_z) + 12.867*(s1.dered_i-s1.dered_z) - 3.8069 between -0.5 and 9.5
AND -0.606*(s2.dered_r-s2.dered_i)*(s2.dered_r-s2.dered_i) + 6.0251*(s2.dered_r-s2.dered_i) - 3.1214 between -0.5 and 9.5
AND -3.1554*(s2.dered_i-s2.dered_z)*(s2.dered_i-s2.dered_z) + 12.867*(s2.dered_i-s2.dered_z) - 3.8069 between -0.5 and 9.5

AND abs((-0.606*(s1.dered_r-s1.dered_i)*(s1.dered_r-s1.dered_i) + 6.0251*(s1.dered_r-s1.dered_i) - 3.1214) —(-
0.606*(s2.dered_r-s2.dered_i)*(s2.dered_r-s2.dered_i) + 6.0251*(s2.dered_r-s2.dered_i) - 3.1214)) < 0.25
```

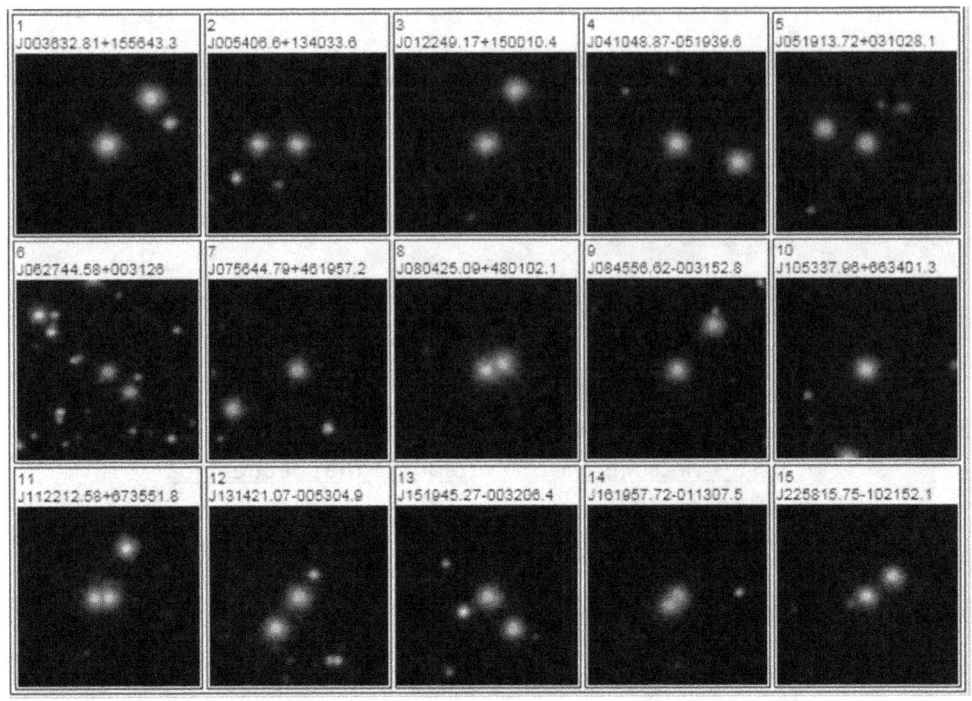

Fig. 55 and Fig.56 – Matching stars by their spectral type

| # | RA | DE | U | G | R | I | Z | RA2 | DE2 | U | G | R | I | Z | DIST | ST1 | ST2 |
|---|----|----|---|---|---|---|---|-----|-----|---|---|---|---|---|------|-----|-----|
| 1 | 9.1367 | 15.9454 | 19.413 | 16.760 | 15.380 | 14.784 | 14.466 | 9.1334 | 15.9489 | 19.501 | 16.872 | 15.491 | 14.862 | 14.531 | 17.01 | 0 | 0 |
| 2 | 13.5275 | 13.6760 | 20.194 | 17.526 | 16.178 | 15.326 | 14.884 | 13.5305 | 13.6760 | 20.181 | 17.638 | 16.243 | 15.361 | 14.899 | 10.26 | 2 | 2 |
| 3 | 20.7049 | 15.0029 | 19.753 | 17.215 | 15.796 | 14.666 | 14.093 | 20.7027 | 15.0069 | 19.789 | 17.253 | 15.857 | 14.709 | 14.127 | 16.28 | 3 | 3 |
| 4 | 62.7036 | -5.3277 | 19.441 | 16.962 | 15.608 | 14.792 | 14.403 | 62.6990 | -5.3291 | 19.419 | 16.999 | 15.640 | 14.835 | 14.428 | 17.29 | 1 | 1 |
| 5 | 79.8072 | 3.1745 | 19.874 | 17.771 | 16.458 | 15.195 | 14.478 | 79.8102 | 3.1756 | 20.282 | 17.827 | 16.503 | 15.207 | 14.487 | 11.69 | 4 | 4 |
| 6 | 96.9358 | 0.5239 | 17.814 | 16.300 | 15.601 | 14.531 | 14.132 | 96.9342 | 0.5224 | 18.097 | 16.308 | 15.616 | 14.552 | 14.157 | 8.00 | 3 | 3 |
| 7 | 119.1866 | 46.3326 | 20.090 | 17.593 | 16.299 | 15.076 | 14.455 | 119.1936 | 46.3297 | 20.107 | 17.670 | 16.377 | 15.143 | 14.519 | 20.28 | 3 | 3 |
| 8 | 121.1046 | 48.0173 | 19.120 | 16.488 | 15.171 | 14.390 | 13.994 | 121.1028 | 48.0177 | 19.212 | 16.549 | 15.229 | 14.426 | 14.014 | 4.60 | 1 | 1 |
| 9 | 131.4860 | -0.5313 | 20.207 | 17.515 | 16.079 | 14.919 | 14.318 | 131.4833 | -0.5280 | 20.292 | 17.557 | 16.117 | 14.948 | 14.336 | 15.40 | 3 | 3 |
| 10 | 163.4082 | 66.5671 | 20.015 | 17.347 | 15.949 | 15.093 | 14.608 | 163.4116 | 66.5603 | 20.095 | 17.477 | 16.033 | 15.137 | 14.616 | 24.74 | 2 | 2 |
| 11 | 170.5524 | 67.5977 | 20.096 | 17.294 | 15.931 | 15.164 | 14.741 | 170.5550 | 67.5977 | 20.201 | 17.409 | 16.049 | 15.248 | 14.825 | 3.57 | 1 | 1 |
| 12 | 198.5878 | -0.8847 | 19.635 | 16.970 | 15.590 | 14.715 | 14.250 | 198.5895 | -0.8870 | 19.804 | 17.085 | 15.692 | 14.808 | 14.356 | 10.19 | 2 | 2 |
| 13 | 229.9387 | -0.5351 | 19.944 | 17.405 | 15.985 | 15.004 | 14.479 | 229.9368 | -0.5375 | 20.179 | 17.463 | 16.025 | 15.030 | 14.508 | 10.84 | 2 | 2 |
| 14 | 244.9905 | -1.2188 | 20.443 | 17.574 | 16.204 | 14.736 | 13.985 | 244.9911 | -1.2194 | 20.338 | 17.592 | 16.282 | 14.877 | 14.154 | 3.17 | 4 | 4 |
| 15 | 344.5656 | -10.3645 | 20.188 | 17.499 | 16.140 | 15.376 | 14.969 | 344.5637 | -10.3630 | 20.295 | 17.530 | 16.160 | 15.370 | 14.959 | 8.72 | 1 | 1 |

## Exercise 3 – The search for celestial twins in data release 10

The SDSS provides photometric information in five different wavebands and it is possible to extrapolate from the SDSS results obtained from previously studied stars to the absolute magnitude, and hence the distance, of previously unstudied stars. An extension of this concept is that if two stars have, essentially, identical re-reddened photometry in all five SDSS wavebands then they must be at the same distance, especially if the angular separation between them is less than 30 arc seconds.

The SQL programme designed to identify "identical twin" star systems is subdivided into sections.

- The first section identifies what data fields are to be extracted from the different tables that are available. The "Neighbors" tables contains information on the SDSS objects within 30 arc second of each other and the "PhotoObjAll" table is a full photometric catalogue for each SDSS object.
- The second section confines the search to primary objects (mode=1), that are stars (type=6), that have clean photometry (clean=1) and an extinction (galactic reddening) in the r band of <0.1 magnitudes.
- The third section is used to confine the experiment to brighter stars.
- The fourth section constrains the reported magnitudes to values that avoid both excessively bright or faint (and so unreliable) values.
- The final section is used to identify those stars where all five listed de-reddened magnitudes (u, g, r, i and z bands) are within 0.05 magnitudes.

```
SELECT TOP 20000 s1.ra as RA, s1.dec as DE, s1.psfMag_u as U, s1.psfMag_g as G, s1.psfMag_r as R, s1.psfMag_i as I,
s1.psfMag_z as Z, s1.extinction_u as UE, s1.extinction_g as GE, s1.extinction_r as RE, s1.extinction_i as IE, s1.extinction_z as ZE,
s2.ra as RA2, s2.dec as DE2, s2.psfMag_u as U2, s2.psfMag_g as G2, s2.psfMag_r as R2, s2.psfMag_i as I2, s2.psfMag_z as Z2,
s2.extinction_u as UE2, s2.extinction_g as GE2, s2.extinction_r as RE2, s2.extinction_i as IE2, s2.extinction_z as ZE2, N.distance
as DI

INTO mydb.xray1
FROM PhotoObjAll S1, Neighbors N, PhotoObjAll S2
WHERE S1.ObjID = N.ObjID
AND S2.ObjID = N.NeighborObjID

AND S1.RA between 0 and 180

AND S1.Mode = 1
AND S2.Mode = 1

AND S1.Type = 6
AND S2.Type = 6

AND S1.Clean = 1
AND S2.Clean = 1

AND s1.extinction_r < 0.1
AND s2.extinction_r < 0.1

AND N.DISTANCE between 0.05 and 0.1667
AND s1.psfMag_r between 14.1 and 16.1

AND  s1.psfMag_u between 12.1 and 21.4
AND  s1.psfMag_g between 14.1 and 22.6
```

AND  s1.psfMag_r between 14.1 and 22.3
AND  s1.psfMag_i between 13.8 and 21.7
AND  s1.psfMag_z between 12.3 and 20.1
AND  s2.psfMag_u between 12.1 and 21.4
AND  s2.psfMag_g between 14.1 and 22.6
AND  s2.psfMag_r between 14.1 and 22.3
AND  s2.psfMag_i between 13.8 and 21.7
AND  s2.psfMag_z between 12.3 and 20.1

AND abs((s1.psfMag_u - s1.extinction_u)-( s2.psfMag_u - s2.extinction_u)) < 0.05
AND abs((s1.psfMag_g - s1.extinction_g)-( s2.psfMag_g - s2.extinction_g)) < 0.05
AND abs((s1.psfMag_r - s1.extinction_r)-( s2.psfMag_r - s2.extinction_r)) < 0.05
AND abs((s1.psfMag_i - s1.extinction_i)-( s2.psfMag_I - s2.extinction_i)) < 0.05
AND abs((s1.psfMag_z - s1.extinction_z)-( s2.psfMag_z - s2.extinction_z)) < 0.05

## RESULTS

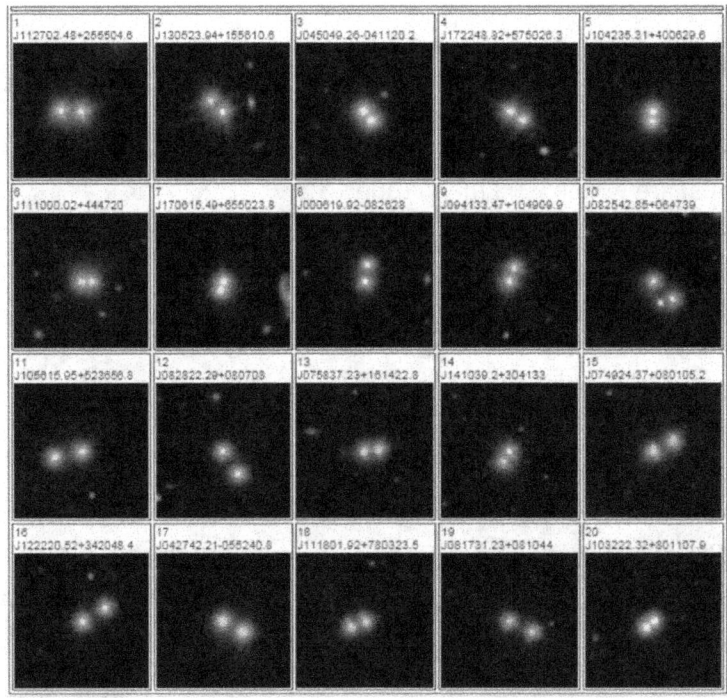

Fig. 57 - Thumbnail images of the twins (48 x 48 arc seconds)

| | DE-REDDENED MAGNITUDE | | | | | DE-REDDENED MAGNITUDE | | | | | RA | | | DECLINATION | | | DETAIL OF DOUBLE STAR | | | |
|---|---|---|---|---|---|---|---|---|---|---|---|---|---|---|---|---|---|---|---|---|
| | Prim u | Prim g | Prim r | Prim i | Priz z | Sec u | Sec g | Sec r | Sec i | Sec z | H | M | S | D | M | S | MAG 1 | MAG2 | SEP | PA |
| 1 | 15.970 | 14.549 | 14.110 | 13.921 | 13.920 | 15.970 | 14.573 | 14.135 | 13.944 | 13.935 | 11 | 27 | 2.49 | 25 | 55 | 4.6 | 14.11 | 14.135 | 7.06 | 87.642 |
| 2 | 16.049 | 14.597 | 14.141 | 13.978 | 13.901 | 16.037 | 14.602 | 14.149 | 13.994 | 13.933 | 13 | 5 | 23.94 | 15 | 56 | 10.65 | 14.141 | 14.149 | 5.862 | 49.839 |
| 3 | 16.016 | 14.769 | 14.421 | 14.313 | 14.315 | 16.020 | 14.768 | 14.436 | 14.335 | 14.343 | 4 | 50 | 49.26 | -4 | 11 | 20.26 | 14.421 | 14.436 | 4.137 | 217.685 |
| 4 | 16.381 | 14.925 | 14.423 | 14.258 | 14.193 | 16.336 | 14.917 | 14.428 | 14.272 | 14.202 | 17 | 22 | 48.83 | 57 | 50 | 26.36 | 14.423 | 14.428 | 5.115 | 235.757 |
| 5 | 16.839 | 15.064 | 14.455 | 14.253 | 14.135 | 16.845 | 15.082 | 14.479 | 14.279 | 14.155 | 10 | 42 | 35.31 | 40 | 6 | 29.7 | 14.455 | 14.479 | 3.482 | 175.706 |
| 6 | 15.960 | 14.866 | 14.463 | 14.367 | 14.360 | 15.987 | 14.881 | 14.479 | 14.384 | 14.374 | 11 | 10 | 0.03 | 44 | 47 | 20.07 | 14.463 | 14.479 | 3.484 | 271.532 |
| 7 | 16.009 | 14.890 | 14.498 | 14.398 | 14.413 | 16.029 | 14.914 | 14.510 | 14.398 | 14.383 | 17 | 6 | 15.5 | 65 | 50 | 23.83 | 14.498 | 14.51 | 3.081 | 162.073 |
| 8 | 16.680 | 15.122 | 14.533 | 14.312 | 14.273 | 16.727 | 15.125 | 14.554 | 14.336 | 14.296 | 0 | 6 | 19.92 | -8 | 26 | 28.03 | 14.533 | 14.554 | 6.071 | 353.166 |
| 9 | 16.036 | 14.879 | 14.585 | 14.447 | 14.463 | 16.036 | 14.902 | 14.613 | 14.485 | 14.510 | 9 | 41 | 33.47 | 10 | 49 | 9.94 | 14.585 | 14.613 | 4.875 | 339.869 |
| 10 | 16.272 | 15.033 | 14.597 | 14.482 | 14.457 | 16.294 | 15.060 | 14.620 | 14.497 | 14.475 | 8 | 25 | 42.86 | 6 | 47 | 39.08 | 14.597 | 14.62 | 9.319 | 227.294 |
| 11 | 16.098 | 15.036 | 14.692 | 14.584 | 14.591 | 16.135 | 15.069 | 14.734 | 14.612 | 14.593 | 10 | 56 | 15.95 | 52 | 36 | 56.9 | 14.692 | 14.734 | 9.689 | 102.233 |
| 12 | 16.215 | 15.040 | 14.693 | 14.619 | 14.546 | 16.256 | 15.078 | 14.733 | 14.665 | 14.593 | 8 | 28 | 22.3 | 8 | 7 | 8.09 | 14.693 | 14.733 | 9.598 | 214.531 |
| 13 | 15.922 | 14.903 | 14.697 | 14.660 | 14.725 | 15.908 | 14.933 | 14.731 | 14.688 | 14.749 | 7 | 58 | 37.23 | 16 | 14 | 22.85 | 14.697 | 14.731 | 4.965 | 277.081 |
| 14 | 16.766 | 15.201 | 14.698 | 14.530 | 14.449 | 16.812 | 15.242 | 14.735 | 14.569 | 14.484 | 14 | 10 | 39.21 | 30 | 41 | 33.05 | 14.698 | 14.735 | 4.093 | 149.808 |
| 15 | 16.103 | 15.008 | 14.742 | 14.639 | 14.672 | 16.094 | 15.027 | 14.745 | 14.628 | 14.646 | 7 | 49 | 24.38 | 8 | 1 | 5.26 | 14.742 | 14.745 | 7.928 | 297.622 |
| 16 | 17.266 | 15.396 | 14.775 | 14.614 | 14.533 | 17.267 | 15.420 | 14.789 | 14.627 | 14.561 | 12 | 22 | 20.53 | 34 | 20 | 48.41 | 14.775 | 14.789 | 9.453 | 301.143 |
| 17 | 18.755 | 16.085 | 14.818 | 14.328 | 14.099 | 18.749 | 16.095 | 14.827 | 14.338 | 14.104 | 4 | 27 | 42.22 | -5 | 52 | 40.86 | 14.818 | 14.827 | 7.972 | 242.169 |
| 18 | 16.489 | 15.255 | 14.890 | 14.792 | 14.804 | 16.484 | 15.275 | 14.917 | 14.813 | 14.808 | 11 | 18 | 1.92 | 78 | 3 | 23.54 | 14.89 | 14.917 | 5.33 | 110.349 |
| 19 | 18.899 | 16.335 | 14.947 | 14.258 | 13.896 | 18.904 | 16.357 | 14.967 | 14.267 | 13.901 | 8 | 17 | 31.24 | 8 | 10 | 44.09 | 14.947 | 14.967 | 8.721 | 245.271 |
| 20 | 16.564 | 15.359 | 15.019 | 14.900 | 14.910 | 16.585 | 15.383 | 15.033 | 14.910 | 14.923 | 10 | 32 | 22.33 | 80 | 11 | 7.98 | 15.019 | 15.033 | 3.411 | 127.875 |

Fig. 58 - A small sample of "5-way celestial twins"

The problem with searching for 5-way pairs is that a large percentage of stars have a u-band magnitude outside the range 12.1 to 21.4. In a random sample of over 2,500 stars nearly 85% had a u-band magnitude fainter than 21.4 which would mean that they would be filtered out by the conditions built into the programme. For that reason 4-way pairs were hunted for by modifying the SQL programme as shown below.

*AND s1.psfMag_u between 12.1 and 21.4 – line deleted*
AND s1.psfMag_g between 14.1 and 22.6
AND s1.psfMag_r between 14.1 and 22.3
AND s1.psfMag_i between 13.8 and 21.7
AND s1.psfMag_z between 12.3 and 20.1
*AND s1.psfMag_u between 12.1 and 21.4 – line deleted*
AND s2.psfMag_g between 14.1 and 22.6
AND s2.psfMag_r between 14.1 and 22.3
AND s2.psfMag_i between 13.8 and 21.7
AND s2.psfMag_z between 12.3 and 20.1

*AND abs((s1.psfMag_u - s1.extinction_u)-( s2.psfMag_u - s2.extinction_u)) < 0.05 – line deleted*
AND abs((s1.psfMag_g - s1.extinction_g)-( s2.psfMag_g - s2.extinction_g)) < 0.05
AND abs((s1.psfMag_r - s1.extinction_r)-( s2.psfMag_r - s2.extinction_r)) < 0.05
AND abs((s1.psfMag_i - s1.extinction_i)-( s2.psfMag_I - s2.extinction_i)) < 0.05
AND abs((s1.psfMag_z - s1.extinction_z)-( s2.psfMag_z - s2.extinction_z)) < 0.05

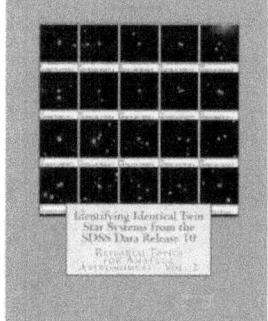

Identifying Identical Twin Stars from SDSS Data Release 10
by Martin Nicholson
The book is available on Amazon (www.amazon.com)

Data mining the SDSS Data Release 10 has yielded a total of 481 pairs of stars separated by between 3 and 30 arc second where all five listed de-reddened magnitudes (u, g, r, i and z bands) for the two stars are within 0.05 magnitudes. 473 of these binary stars are believed to be new discoveries. The SDSS provides photometry in five different wavebands and these results can be used to estimate the spectral energy distribution of the star. It is possible to extrapolate from the SDSS results obtained from previously studied stars to the absolute magnitude, and hence the distance, of previously unstudied stars. An extension of this idea of photometric parallax is the idea, presented for the first time in this paper, that if two stars have, essentially, identical re-reddened photometry in all five SDSS wavebands then they must be at the same distance, especially if the angular separation between them is less than 30 arc seconds.

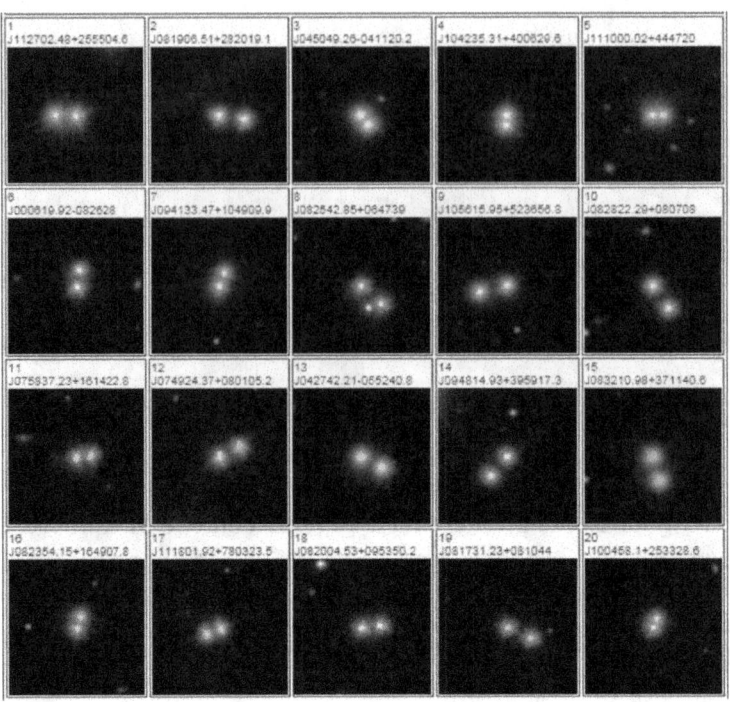

Fig. 58 and Fig. 59 - A sample of "4-way celestial twins"
Thumbnail images are 48 x 48 arc seconds

| | DE-REDDENED MAGNITUDE | | | | | DE-REDDENED MAGNITUDE | | | | | RA | | | DECLINATION | | | DETAIL OF DOUBLE STAR | | | |
|---|---|---|---|---|---|---|---|---|---|---|---|---|---|---|---|---|---|---|---|---|
| | Prim u | Prim g | Prim r | Prim i | Priz z | Sec u | Sec g | Sec r | Sec i | Sec z | H | M | S | D | M | S | MAG 1 | MAG2 | SEP | PA |
| 1 | 15.970 | 14.549 | 14.110 | 13.921 | 13.920 | 15.970 | 14.573 | 14.135 | 13.944 | 13.935 | 11 | 27 | 2.49 | 25 | 55 | 4.6 | 14.11 | 14.135 | 7.06 | 87.642 |
| 2 | 16.693 | 15.022 | 14.415 | 14.207 | 14.109 | 16.827 | 15.068 | 14.428 | 14.217 | 14.117 | 8 | 19 | 6.52 | 28 | 20 | 19.18 | 14.415 | 14.428 | 9.127 | 261.338 |
| 3 | 16.016 | 14.769 | 14.421 | 14.313 | 14.315 | 16.020 | 14.768 | 14.436 | 14.335 | 14.343 | 4 | 50 | 49.26 | -4 | 11 | 20.26 | 14.421 | 14.436 | 4.137 | 217.685 |
| 4 | 16.839 | 15.064 | 14.455 | 14.253 | 14.135 | 16.845 | 15.082 | 14.479 | 14.279 | 14.155 | 10 | 42 | 35.31 | 40 | 6 | 29.7 | 14.455 | 14.479 | 3.482 | 175.706 |
| 5 | 15.960 | 14.866 | 14.463 | 14.367 | 14.360 | 15.987 | 14.881 | 14.479 | 14.384 | 14.374 | 11 | 10 | 0.03 | 44 | 47 | 20.07 | 14.463 | 14.479 | 3.484 | 271.532 |
| 6 | 16.680 | 15.122 | 14.533 | 14.312 | 14.273 | 16.727 | 15.125 | 14.554 | 14.336 | 14.296 | 0 | 6 | 19.92 | -8 | 26 | 28.03 | 14.533 | 14.554 | 6.071 | 353.166 |
| 7 | 16.036 | 14.879 | 14.585 | 14.447 | 14.463 | 16.036 | 14.902 | 14.613 | 14.485 | 14.510 | 9 | 41 | 33.47 | 10 | 49 | 9.94 | 14.585 | 14.613 | 4.875 | 339.869 |
| 8 | 16.272 | 15.033 | 14.597 | 14.482 | 14.457 | 16.294 | 15.060 | 14.620 | 14.497 | 14.475 | 8 | 25 | 42.86 | 6 | 47 | 39.08 | 14.597 | 14.62 | 9.319 | 227.294 |
| 9 | 16.098 | 15.036 | 14.692 | 14.584 | 14.591 | 16.135 | 15.069 | 14.734 | 14.612 | 14.612 | 10 | 56 | 15.95 | 52 | 36 | 56.9 | 14.692 | 14.734 | 9.689 | 102.233 |
| 10 | 16.215 | 15.040 | 14.693 | 14.619 | 14.546 | 16.256 | 15.078 | 14.733 | 14.665 | 14.593 | 8 | 28 | 22.3 | 8 | 7 | 8.09 | 14.693 | 14.733 | 9.598 | 214.531 |
| 11 | 15.922 | 14.903 | 14.697 | 14.660 | 14.725 | 15.908 | 14.933 | 14.731 | 14.686 | 14.749 | 7 | 58 | 37.23 | 16 | 14 | 22.85 | 14.697 | 14.731 | 4.965 | 277.081 |
| 12 | 16.103 | 15.008 | 14.742 | 14.639 | 14.672 | 16.094 | 15.027 | 14.745 | 14.628 | 14.646 | 7 | 49 | 24.38 | 8 | 1 | 5.26 | 14.742 | 14.745 | 7.928 | 297.622 |
| 13 | 18.755 | 16.085 | 14.818 | 14.328 | 14.099 | 18.749 | 16.095 | 14.827 | 14.338 | 14.104 | 4 | 27 | 42.22 | -5 | 52 | 40.86 | 14.818 | 14.827 | 7.972 | 242.169 |
| 14 | 16.592 | 15.284 | 14.819 | 14.670 | 14.650 | 16.486 | 15.257 | 14.826 | 14.675 | 14.662 | 9 | 48 | 14.93 | 39 | 59 | 17.36 | 14.819 | 14.826 | 8.815 | 139.227 |
| 15 | 18.902 | 16.280 | 14.826 | 13.778 | 13.235 | 18.978 | 16.311 | 14.866 | 13.810 | 13.232 | 8 | 32 | 10.98 | 37 | 11 | 40.66 | 14.826 | 14.866 | 8.677 | 195.615 |
| 16 | 16.484 | 15.214 | 14.851 | 14.771 | 14.765 | 16.324 | 15.211 | 14.884 | 14.793 | 14.781 | 8 | 23 | 54.16 | 16 | 49 | 7.88 | 14.851 | 14.884 | 4.229 | 344.543 |
| 17 | 16.489 | 15.255 | 14.890 | 14.792 | 14.804 | 16.484 | 15.275 | 14.917 | 14.813 | 14.808 | 11 | 18 | 1.92 | 78 | 3 | 23.54 | 14.89 | 14.917 | 5.33 | 110.349 |
| 18 | 16.738 | 15.402 | 14.945 | 14.808 | 14.792 | 16.798 | 15.444 | 14.980 | 14.847 | 14.829 | 8 | 20 | 4.53 | 9 | 53 | 50.25 | 14.945 | 14.98 | 6.347 | 277.433 |
| 19 | 18.899 | 16.335 | 14.947 | 14.258 | 13.896 | 18.904 | 16.357 | 14.967 | 14.267 | 13.901 | 8 | 17 | 31.24 | 8 | 10 | 44.09 | 14.947 | 14.967 | 8.721 | 245.271 |
| 20 | 17.400 | 15.638 | 14.990 | 14.799 | 14.727 | 17.468 | 15.688 | 15.030 | 14.844 | 14.756 | 10 | 4 | 58.1 | 25 | 33 | 28.64 | 14.99 | 15.03 | 3.164 | 340.291 |

# Data mining for uncatalogued binary and double star systems in UCAC4

There are a number of practical and scientific reasons why this is probably no longer a viable project.

#1 – UCAC4, plus its earlier versions UCAC 2 and 3, have been examined in detail by a number of different researchers. The two largest studies, in terms of discovery claims, were by Greaves in 2004 and by Zacharias et al in 2013 and I suspect that there are few, if any, noteworthy pairs remaining to be discovered that are not already in the public domain.

Any astronomical data miner determined to re-examine this catalogue should concentrate on wider pairs and/or pairs where both components have a total proper motion in the range 40 to 60 mas/yr.

| # | PRIMARY RA | PRIMARY DEC | SECONDARY RA | SECONDARY DEC | PRIMARY PM in RA | PRIMARY PM in DEC | SECONDARY PM in RA | SECONDARY PM in DEC | PRIMARY MAGNITUDE | SECONDARY MAGNITUDE | CPM Pair SEP (arc sec) | CPM Pair PA (degrees) |
|----|----------|-----------|-----------|-----------|------|------|------|------|--------|--------|------|-----|
| 1 | 6.3926 | 82.8180 | 6.2602 | 82.8279 | 45.9 | -24.0 | 47.3 | -25.4 | 11.857 | 13.158 | 16.8 | 193 |
| 2 | 7.2148 | -30.9165 | 7.2334 | -30.9304 | -61.5 | -22.2 | -64.0 | -24.8 | 11.200 | 11.977 | 76.3 | 131 |
| 3 | 10.5641 | -8.5508 | 10.5739 | -8.5655 | -17.8 | 51.9 | -17.8 | 48.8 | 9.587 | 10.392 | 63.4 | 146 |
| 4 | 11.3686 | 50.7710 | 11.3401 | 50.7858 | 50.3 | -13.7 | 48.3 | -13.1 | 10.382 | 11.743 | 29.6 | 128 |
| 5 | 11.9004 | -19.5897 | 11.8879 | -19.6112 | 45.5 | 33.8 | 44.8 | 33.6 | 11.937 | 12.871 | 88.0 | 209 |
| 6 | 12.6520 | -26.0349 | 12.6578 | -26.0320 | 69.7 | 52.8 | 69.6 | 60.9 | 10.440 | 12.910 | 21.4 | 61 |
| 7 | 12.7774 | -7.5342 | 12.7797 | -7.5324 | 47.6 | 9.3 | 54.0 | 9.9 | 11.026 | 13.995 | 10.5 | 53 |
| 8 | 14.8679 | -83.1384 | 14.8458 | -83.1414 | 57.5 | 65.7 | 54.2 | 60.4 | 9.239 | 13.888 | 14.6 | 221 |
| 9 | 17.3261 | 72.4285 | 17.3227 | 72.4239 | 51.6 | -15.5 | 51.0 | -16.0 | 11.359 | 12.011 | 27.0 | 91 |
| 10 | 17.9139 | 79.1656 | 17.8781 | 79.1890 | 48.1 | -21.6 | 48.3 | -25.0 | 11.084 | 13.833 | 71.0 | 264 |
| 11 | 18.0830 | -10.8668 | 18.0632 | -10.8650 | -43.8 | -36.1 | -42.3 | -34.3 | 13.248 | 13.664 | 70.4 | 275 |
| 12 | 22.5444 | -50.3934 | 22.5406 | -50.3900 | -18.9 | -51.5 | -19.6 | -45.9 | 11.673 | 11.734 | 15.0 | 324 |
| 13 | 23.7861 | -23.4475 | 23.7632 | -23.4600 | 48.5 | 61.1 | 48.6 | 59.7 | 8.367 | 10.009 | 88.1 | 239 |
| 14 | 27.2801 | 82.7817 | 27.3408 | 82.7961 | 60.0 | 15.5 | 60.4 | 17.1 | 13.078 | 13.304 | 34.8 | 288 |
| 15 | 28.6519 | -43.1490 | 28.6761 | -43.1468 | -24.4 | -45.1 | -26.9 | -41.0 | 10.893 | 12.217 | 63.9 | 83 |
| 16 | 30.4555 | -22.2867 | 30.4525 | -22.2836 | -22.2 | -42.7 | -24.6 | -41.0 | 9.805 | 13.513 | 14.9 | 318 |
| 17 | 31.2496 | -19.6264 | 31.2707 | -19.6297 | 61.2 | 30.2 | 60.1 | 28.6 | 11.910 | 12.912 | 72.3 | 100 |
| 18 | 31.8843 | 55.7488 | 31.8976 | 55.7488 | 49.9 | -22.1 | 51.2 | -21.8 | 13.041 | 13.489 | 41.2 | 331 |
| 19 | 36.0969 | -17.8196 | 36.1104 | -17.8024 | 43.5 | 7.3 | 42.5 | 7.1 | 9.484 | 12.677 | 77.3 | 37 |
| 20 | 38.2328 | -41.8605 | 38.2382 | -41.8837 | 38.3 | 59.8 | 37.8 | 60.4 | 12.599 | 13.647 | 84.6 | 170 |
| 21 | 38.2421 | -43.2598 | 38.2344 | -43.2777 | 44.7 | -3.2 | 42.8 | -3.7 | 11.193 | 12.601 | 67.3 | 198 |
| 22 | 38.4796 | -14.0526 | 38.4724 | -14.0381 | 49.9 | -18.1 | 50.2 | -18.3 | 9.948 | 10.427 | 58.0 | 334 |
| 23 | 41.3659 | 50.2549 | 41.3364 | 50.2639 | -53.3 | -31.8 | -58.8 | -32.6 | 10.683 | 13.929 | 75.3 | 296 |
| 24 | 42.7096 | -75.6665 | 42.6982 | -75.6715 | 47.4 | 17.1 | 46.5 | 17.4 | 12.192 | 13.506 | 20.7 | 210 |
| 25 | 53.6006 | -7.8687 | 53.5972 | -7.8540 | 41.2 | 24.4 | 40.3 | 23.9 | 12.083 | 12.930 | 54.4 | 347 |
| 26 | 55.0852 | -31.3636 | 55.0818 | -31.3574 | -18.0 | -43.3 | -15.9 | -42.6 | 12.093 | 12.991 | 24.5 | 335 |
| 27 | 57.4109 | -75.1788 | 57.3429 | -75.1659 | 66.5 | 24.0 | 60.9 | 23.7 | 11.907 | 13.612 | 77.8 | 306 |
| 28 | 58.7864 | 61.9449 | 58.7714 | 61.9470 | 9.1 | 46.5 | 8.5 | 45.8 | 11.263 | 11.546 | 26.6 | 287 |
| 29 | 61.8077 | -67.5536 | 61.8188 | -67.5488 | 14.3 | 46.2 | 13.9 | 44.0 | 11.777 | 12.975 | 23.1 | 41 |
| 30 | 66.0759 | -84.4258 | 66.1540 | -84.4124 | 21.6 | 43.8 | 20.1 | 46.1 | 11.454 | 12.383 | 55.4 | 30 |
| 31 | 70.3302 | -53.6725 | 70.3428 | -53.6865 | 43.8 | -21.6 | 40.5 | -21.5 | 12.776 | 13.489 | 57.0 | 152 |
| 32 | 71.2805 | -51.0207 | 71.3085 | -51.0110 | 6.4 | 54.4 | 6.5 | 55.3 | 11.797 | 13.400 | 72.4 | 61 |
| 33 | 79.5815 | 49.3378 | 79.5920 | 49.3585 | 47.8 | -56.7 | 47.7 | -46.7 | 12.785 | 13.744 | 15.1 | 192 |
| 34 | 79.8904 | -42.8839 | 79.8845 | -42.8888 | -6.4 | 47.7 | -6.8 | 43.6 | 12.227 | 12.416 | 23.6 | 221 |
| 35 | 80.8147 | -57.8501 | 80.8126 | -57.8551 | 17.3 | 104.7 | 16.6 | 101.3 | 11.575 | 13.496 | 18.6 | 192 |
| 36 | 83.2472 | 75.4350 | 83.2343 | 75.4415 | -51.5 | 47.2 | -53.5 | 47.4 | 11.148 | 12.716 | 26.3 | 334 |
| 37 | 85.2867 | -60.0014 | 85.2952 | -59.9851 | 7.8 | 57.1 | 8.9 | 56.0 | 11.637 | 11.814 | 60.8 | 15 |
| 38 | 86.4276 | -45.0295 | 86.4179 | -45.0114 | 11.7 | 57.4 | 10.6 | 60.0 | 12.445 | 13.077 | 69.7 | 339 |
| 39 | 104.4967 | -41.4780 | 104.4723 | -41.4856 | -16.2 | 57.8 | -17.7 | 51.7 | 9.512 | 12.209 | 71.3 | 248 |
| 40 | 106.3400 | 73.9275 | 106.2793 | 73.9238 | -42.2 | -31.6 | -41.8 | -34.5 | 11.552 | 13.614 | 62.0 | 258 |

Fig. 60 - Candidate Common Proper Motion Pairs in UCAC4 – Part 1

#2 – Astronomers using the UCAC4 catalogue will almost inevitably find themselves unwilling participants in the current scientific controversy surrounding this catalogue.

Greaves was, with hindsight, far too liberal in his definition of what constituted a common proper motion pair. Some years ago he wrote, *"Some I've published in the past are only candidates. Some aren't even pairs. About a third seem to be certainly real, a third unconfirmed, i.e. candidates, and the remaining third totally spurious"*. It is unclear why so many of these "totally spurious" binary star systems remain in the catalogue.

> New Northern hemisphere common proper motion pairs
> Monthly Notices of the Royal Astronomical Society, Volume 355, Issue 2, pp. 585-590
> John Greaves

| # | PRIMARY RA | DEC | SECONDARY RA | DEC | PRIMARY PM in RA | PM in DEC | SECONDARY PM in RA | PM in DEC | PRIMARY MAGNITUDE | SECONDARY MAGNITUDE | CPM Pair SEP (arc sec) | PA (degrees) |
|---|---|---|---|---|---|---|---|---|---|---|---|---|
| 41 | 113.7538 | -47.1506 | 113.7563 | -47.1621 | -35.8 | 45.7 | -39.2 | 48.5 | 11.533 | 12.776 | 41.8 | 171 |
| 42 | 114.3642 | -74.9815 | 114.3153 | -74.9654 | -17.4 | 54.9 | -16.2 | 52.4 | 10.404 | 13.301 | 73.8 | 322 |
| 43 | 114.8853 | -45.1140 | 114.8868 | -45.1095 | -26.5 | 63.7 | -23.7 | 61.5 | 12.287 | 13.747 | 16.6 | 13 |
| 44 | 115.8225 | -25.0616 | 115.8192 | -25.0568 | -56.8 | 37.8 | -67.5 | 37.8 | 10.748 | 13.291 | 20.4 | 327 |
| 45 | 117.2198 | -41.0799 | 117.2505 | -41.0868 | -11.6 | 77.5 | -11.4 | 74.6 | 12.127 | 13.685 | 87.1 | 107 |
| 46 | 122.3202 | -61.0557 | 122.2832 | -61.0636 | 57.8 | -105.6 | 53.5 | -109.8 | 8.905 | 12.482 | 70.4 | 246 |
| 47 | 122.3224 | -47.7983 | 122.3090 | -47.7949 | -6.6 | 48.7 | -6.6 | 54.0 | 10.587 | 13.950 | 34.7 | 291 |
| 48 | 123.8373 | 66.3081 | 123.7977 | 66.3068 | -43.4 | -10.3 | -41.7 | -10.4 | 10.072 | 11.373 | 57.4 | 265 |
| 49 | 125.2975 | -53.8524 | 125.2735 | -53.8718 | -43.5 | 45.5 | -44.5 | 45.4 | 11.506 | 11.838 | 86.4 | 216 |
| 50 | 128.6942 | -63.2839 | 128.6954 | -63.2807 | -25.2 | 44.6 | -24.3 | 46.1 | 13.289 | 13.379 | 11.7 | 10 |
| 51 | 129.4450 | -32.4648 | 129.4430 | -32.4660 | 45.7 | -41.8 | 40.4 | -42.6 | 10.043 | 12.366 | 7.5 | 236 |
| 52 | 130.0695 | -56.7762 | 130.0796 | -56.7955 | -41.8 | 59.3 | -45.8 | 60.2 | 8.921 | 12.491 | 72.1 | 164 |
| 53 | 131.5652 | -28.6582 | 131.5729 | -28.6609 | -43.5 | 25.9 | -48.8 | 24.0 | 12.686 | 13.467 | 26.3 | 112 |
| 54 | 133.8730 | -31.2364 | 133.8817 | -31.2316 | 9.4 | 71.6 | 11.3 | 70.8 | 11.535 | 13.135 | 31.9 | 57 |
| 55 | 134.5029 | -58.3659 | 134.5132 | -58.3705 | -44.6 | 22.5 | -44.6 | 26.0 | 9.059 | 12.102 | 25.8 | 131 |
| 56 | 134.8137 | -69.3118 | 134.8069 | -69.3290 | -27.8 | 53.4 | -24.9 | 54.7 | 13.116 | 13.169 | 62.3 | 188 |
| 57 | 134.9460 | -20.8105 | 134.9235 | -20.7986 | -41.2 | 25.1 | -41.5 | 22.0 | 8.080 | 10.867 | 86.9 | 300 |
| 58 | 135.5823 | 61.6609 | 135.5781 | 61.6572 | -40.1 | -76.1 | -40.2 | -79.6 | 11.899 | 12.150 | 15.0 | 209 |
| 59 | 138.8921 | -8.5539 | 138.9136 | -8.5632 | -49.7 | 27.2 | -52.6 | 24.5 | 11.862 | 13.913 | 83.5 | 114 |
| 60 | 146.9523 | -51.4408 | 146.9567 | -51.4387 | -43.7 | 38.2 | -40.0 | 41.2 | 12.080 | 13.391 | 12.5 | 52 |
| 61 | 150.7035 | -42.4168 | 150.7353 | -42.4248 | -40.5 | 31.6 | -43.4 | 35.7 | 9.476 | 11.466 | 89.5 | 109 |
| 62 | 154.8121 | -87.9877 | 154.8815 | -88.0088 | -44.6 | 22.4 | -44.5 | 24.0 | 12.467 | 12.828 | 76.4 | 173 |
| 63 | 155.6532 | -58.8010 | 155.6612 | -58.8000 | -47.5 | 20.8 | -48.1 | 23.8 | 11.758 | 13.387 | 15.2 | 76 |
| 64 | 155.6612 | -25.8402 | 155.6619 | -25.8418 | -129.0 | 101.0 | -136.0 | 102.0 | 12.685 | 13.218 | 6.1 | 158 |
| 65 | 157.9452 | -63.9520 | 157.9847 | -63.9580 | -51.7 | 25.2 | -51.2 | 23.3 | 10.946 | 12.235 | 66.1 | 109 |
| 66 | 158.2063 | -41.4884 | 158.2015 | -41.4880 | -49.7 | -25.8 | -47.8 | -22.2 | 10.886 | 13.187 | 12.9 | 277 |
| 67 | 159.6153 | -62.3570 | 159.6132 | -62.3672 | 51.8 | -57.6 | 44.8 | -58.5 | 10.896 | 13.725 | 36.7 | 186 |
| 68 | 160.4398 | -50.9955 | 160.4362 | -50.9952 | -98.8 | 48.4 | -106.7 | 50.7 | 11.311 | 12.386 | 8.2 | 276 |
| 69 | 162.1325 | -71.4984 | 162.1400 | -71.5043 | -58.6 | 51.9 | -58.9 | 51.9 | 12.585 | 12.874 | 22.8 | 158 |
| 70 | 164.6460 | -43.7293 | 164.6359 | -43.7370 | -34.9 | -62.9 | -38.6 | -62.4 | 12.292 | 13.854 | 38.2 | 224 |
| 71 | 171.4428 | -21.6068 | 171.4411 | -21.6021 | 43.4 | -28.5 | 41.9 | -24.6 | 12.110 | 12.204 | 17.9 | 341 |
| 72 | 173.9879 | -60.7172 | 173.9888 | -60.7380 | -26.6 | -43.4 | -26.2 | -51.5 | 11.895 | 12.525 | 74.8 | 179 |
| 73 | 175.2224 | -55.6764 | 175.2085 | -55.6958 | -48.2 | 35.7 | -47.7 | 30.8 | 12.171 | 13.208 | 75.2 | 202 |
| 74 | 177.4990 | -55.8502 | 177.4970 | -55.8576 | 60.5 | -22.2 | 61.2 | -24.3 | 10.912 | 11.178 | 27.1 | 189 |
| 75 | 179.1309 | -43.7277 | 179.1375 | -43.7248 | -25.5 | 40.6 | -28.2 | 41.0 | 11.840 | 11.918 | 19.9 | 59 |
| 76 | 180.0373 | -67.7774 | 180.0677 | -67.7700 | -21.4 | -46.4 | -21.5 | -47.2 | 9.032 | 12.935 | 49.3 | 57 |
| 77 | 188.3384 | -15.3300 | 188.3474 | -15.3287 | -63.5 | -83.9 | -60.2 | -76.7 | 10.664 | 13.395 | 31.8 | 82 |
| 78 | 195.0985 | -12.6033 | 195.1021 | -12.6058 | -16.5 | -42.6 | -15.7 | -43.9 | 11.435 | 11.490 | 15.6 | 125 |
| 79 | 200.6652 | -31.4372 | 200.6879 | -31.4436 | -59.8 | 42.9 | -58.8 | 44.5 | 10.550 | 12.559 | 73.6 | 108 |
| 80 | 201.4288 | -63.4701 | 201.4564 | -63.4579 | -20.6 | -43.2 | -22.9 | -41.7 | 13.785 | 13.835 | 62.6 | 45 |

Fig. 61 - Candidate Common Proper Motion Pairs in UCAC4 – Part 2

The 2013 paper by Zacharias et al contains some results first published by amateur astronomers several years previously. The amateur results were shared at the time with the professional astronomers at USNO and at the time of writing (May 2014) it still has not been explained why the standard acknowledgement to this earlier work wasn't made.

---

THE FOURTH US NAVAL OBSERVATORY CCD ASTROGRAPH CATALOG (UCAC4)
http://iopscience.iop.org/1538-3881/145/2/44
Zacharius et al (2013)

---

#3 – For many years one of the few perks associated with double star astronomy was that the standard catalogue identified the <u>discoverer</u> of each system. The curious – and almost entirely unpublicised – decision by USNO staff to move away from this system to one where any common proper motion pairs discovered using UCAC4 data are automatically credited to the USNO must be resisted.

| Full | RAJ2000 "h:m:s" | DEJ2000 "d:m:s" | Disc | Comp | Obs1 yr | Obs2 yr | Nobs | pa1 deg | pa2 deg | sep1 arcsec | sep2 arcsec | mag1 mag | mag2 mag | Notes |
|---|---|---|---|---|---|---|---|---|---|---|---|---|---|---|
| 1 | 00 40 06.26 | +50 14 15.5 | NI 2 | | 1998 | 2000 | 2 | 311 | 311 | 24.9 | 24.90 | 10.90 | 13.00 | NVD |
| 2 | 01 41 51.28 | +47 46 25.8 | NI 3 | | 1999 | 2000 | 2 | 359 | 359 | 3.4 | 3.40 | 12.40 | 12.40 | VD |
| 3 | 03 02 09.81 | +26 00 47.6 | NI 5 | | 1997 | 2000 | 2 | 356 | 356 | 11.1 | 11.10 | 13.10 | 13.10 | VD |
| 4 | 03 27 52.36 | +14 50 49.7 | NI 6 | | 1997 | 2000 | 2 | 299 | 299 | 3.7 | 3.70 | 12.80 | 12.80 | VD |
| 5 | 03 44 48.90 | +57 01 41.6 | NI 8 | | 1911 | 2000 | 7 | 314 | 316 | 16.2 | 16.60 | 11.30 | 11.50 | VD |
| 6 | 04 07 57.53 | +04 44 37.8 | NI 9 | | 2000 | 2000 | 2 | 288 | 288 | 2.5 | 2.50 | 13.50 | 13.90 | VD |
| 7 | 04 10 38.31 | +20 02 25.9 | NI 10 | | 1997 | 2000 | 2 | 227 | 227 | 3.3 | 3.30 | 12.30 | 12.30 | VD |
| 8 | 04 12 48.86 | +19 53 52.3 | NI 11 | | 1997 | 2000 | 2 | 228 | 229 | 5.1 | 5.00 | 10.30 | 10.50 | VD |
| 9 | 04 45 25.41 | +29 55 28.4 | NI 12 | | 1901 | 2009 | 9 | 144 | 145 | 84.0 | 84.10 | 11.39 | 11.78 | NVD |
| 10 | 06 20 53.28 | +54 24 59.6 | NI 13 | | 1999 | 2000 | 2 | 139 | 139 | 78.4 | 78.50 | 9.60 | 12.90 | VD |
| 11 | 07 26 40.04 | +26 58 51.5 | NI 14 | | 2000 | 2000 | 2 | 151 | 152 | 5.4 | 5.40 | 10.90 | 11.00 | VD |
| 12 | 07 31 36.14 | +62 01 11.5 | NI 15 | | 1999 | 2000 | 2 | 75 | 75 | 22.9 | 22.90 | 12.00 | 13.50 | VD |
| 13 | 07 48 07.51 | +50 13 04.5 | NI 17 | | 1901 | 2009 | 8 | 341 | 342 | 31.0 | 31.20 | 11.20 | 11.25 | NVD |
| 14 | 08 26 45.45 | +32 50 00.0 | NI 20 | | 1999 | 2000 | 2 | 3 | 1 | 2.5 | 2.70 | 10.80 | 10.90 | VD |
| 15 | 08 45 53.45 | -31 07 19.4 | NI 21 | AB | 1998 | 2000 | 2 | 302 | 302 | 21.1 | 21.00 | 11.10 | 12.00 | VD |
| 16 | 08 52 58.17 | +29 31 44.5 | NI 22 | | 1998 | 2000 | 2 | 157 | 156 | 4.0 | 4.10 | 11.10 | 11.20 | VD |
| 17 | 08 57 42.11 | +55 22 00.1 | NI 23 | | 1999 | 2000 | 2 | 259 | 259 | 12.2 | 12.20 | 13.60 | 13.70 | VD |
| 18 | 10 38 40.78 | +11 32 22.1 | NI 24 | | 2000 | 2000 | 2 | 123 | 123 | 7.9 | 7.90 | 13.20 | 13.70 | VD |
| 19 | 11 22 44.74 | +30 17 40.5 | NI 25 | | 1998 | 2000 | 2 | 85 | 85 | 4.5 | 4.50 | 12.00 | 12.50 | VD |
| 20 | 11 43 53.14 | +33 18 30.6 | NI 27 | | 1998 | 2000 | 2 | 223 | 223 | 3.4 | 3.40 | 13.40 | 13.40 | VD |
| 21 | 11 55 36.20 | +73 30 19.1 | NI 28 | | 2000 | 2000 | 2 | 151 | 151 | 3.3 | 3.30 | 12.60 | 12.70 | VD |
| 22 | 12 52 16.15 | +38 35 40.0 | NI 30 | | 2000 | 2000 | 2 | 158 | 158 | 10.6 | 10.60 | 13.80 | 13.90 | VD |
| 23 | 13 24 29.41 | +41 12 00.8 | NI 31 | | 1998 | 2000 | 2 | 173 | 173 | 8.3 | 8.30 | 11.10 | 11.60 | VD |
| 24 | 15 10 36.61 | +39 23 12.7 | NI 34 | | 1998 | 2000 | 2 | 51 | 54 | 8.6 | 8.30 | 13.00 | 13.30 | VD |
| 25 | 17 15 27.74 | +30 52 36.6 | NI 36 | | 1998 | 2000 | 2 | 166 | 166 | 15.0 | 15.00 | 11.30 | 13.60 | VD |
| 26 | 18 26 24.59 | +11 20 57.4 | NI 38 | | 2000 | 2000 | 1 | 190 | 190 | 7.8 | 7.80 | 13.10 | 13.10 | V |
| 27 | 22 04 17.31 | +09 51 34.6 | NI 43 | | 2000 | 2000 | 2 | 135 | 135 | 40.4 | 40.50 | 11.50 | 13.40 | VD |
| 28 | 22 05 45.36 | +65 38 55.5 | NI 44 | | 2000 | 2000 | 1 | 136 | 136 | 6.8 | 6.80 | 11.90 | 15.40 | V |
| 29 | 23 27 46.08 | +12 23 40.9 | NI 48 | | 1997 | 2000 | 2 | 278 | 279 | 11.6 | 11.60 | 12.50 | 13.80 | VD |
| 30 | 23 50 02.80 | +05 30 46.0 | NI 49 | | 2000 | 2000 | 2 | 125 | 125 | 11.2 | 11.10 | 9.60 | 10.90 | VD |

Fig. 62 - A selection of my discoveries made under the old regime

#4 – Although the UCAC4 is advertised as having, "… precise five-band photometry …" it doesn't have accurate five-band photometry so in a sense any precision is wasted.

In August 2012 (http://www.aavso.org/ucac4-and-apass) I wrote, "The newly released UCAC4 quotes APASS magnitudes (B V g r and i) for large numbers of stars. But lots of the quoted magnitudes are brighter than magnitude 10 and I thought from the APASS documentation that this means they would be saturated. Can they still be used with confidence?"

The answer I was given was, "No they cannot."

| | PRIMARY | | SECONDARY | | PRIMARY | | SECONDARY | | PRIMARY | SECONDARY | CPM Pair | |
|---|---|---|---|---|---|---|---|---|---|---|---|---|
| # | RA | DEC | RA | DEC | PM in RA | PM in DEC | PM in RA | PM in DEC | MAGNITUDE | MAGNITUDE | SEP (arc sec) | PA (degrees) |
| 81 | 203.0550 | 66.1182 | 203.0611 | 66.1198 | -45.0 | 9.5 | -43.2 | 9.8 | 10.506 | 11.943 | 10.6 | 57 |
| 82 | 203.1966 | -56.9376 | 203.1925 | -56.9293 | -50.5 | -62.2 | -48.4 | -65.1 | 11.650 | 12.471 | 30.8 | 345 |
| 83 | 216.1949 | -52.1100 | 216.1822 | -52.1001 | -80.3 | -43.0 | -68.1 | -43.6 | 10.138 | 12.144 | 45.4 | 322 |
| 84 | 218.0821 | -57.5115 | 218.0833 | -57.5163 | -54.1 | 52.0 | -57.1 | 49.8 | 10.814 | 10.916 | 17.4 | 172 |
| 85 | 218.0867 | -46.9014 | 218.1003 | -46.9025 | -56.4 | -41.4 | -62.4 | -41.2 | 10.092 | 11.356 | 33.7 | 97 |
| 86 | 220.9154 | -41.7558 | 220.9016 | -41.7379 | -24.5 | -52.0 | -22.0 | -50.7 | 13.699 | 13.668 | 74.5 | 150 |
| 87 | 223.1981 | -24.7150 | 223.2182 | -24.6991 | -72.4 | -53.8 | -72.5 | -48.0 | 13.703 | 13.708 | 87.2 | 49 |
| 88 | 225.9396 | -47.4698 | 225.9703 | -47.4632 | 43.6 | 50.9 | 47.0 | 57.1 | 13.508 | 13.705 | 78.6 | 72 |
| 89 | 227.3282 | 66.6601 | 227.2764 | 66.6679 | -60.8 | -75.0 | -62.2 | -74.9 | 9.412 | 12.272 | 79.1 | 291 |
| 90 | 232.3069 | 59.4446 | 232.3124 | 59.4228 | -62.6 | 24.0 | -62.2 | 21.6 | 10.985 | 11.839 | 79.0 | 173 |
| 91 | 238.7073 | -56.6457 | 238.7384 | -56.6456 | -57.7 | -75.9 | -51.7 | -83.0 | 13.114 | 13.208 | 61.4 | 90 |
| 92 | 241.1609 | -84.7343 | 241.1683 | -84.7423 | 7.8 | 47.2 | 8.1 | 44.7 | 10.494 | 12.347 | 28.8 | 175 |
| 93 | 243.2229 | -34.4877 | 243.2284 | -34.4863 | -38.3 | -69.4 | -35.7 | -69.5 | 9.469 | 11.418 | 16.8 | 73 |
| 94 | 243.2892 | -65.6007 | 243.2321 | -65.6062 | -49.6 | -124.9 | -47.2 | -120.0 | 13.890 | 13.895 | 87.3 | 257 |
| 95 | 244.0392 | -55.8110 | 244.0157 | -55.8120 | -45.0 | -40.3 | -47.8 | -40.8 | 12.909 | 13.953 | 47.7 | 266 |
| 96 | 246.8191 | -50.7667 | 246.8120 | -50.7647 | -44.5 | -64.1 | -41.9 | -58.3 | 12.956 | 13.512 | 17.7 | 294 |
| 97 | 246.8798 | -54.0683 | 246.8884 | -54.0714 | -39.7 | -84.8 | -45.0 | -81.5 | 9.864 | 13.940 | 21.3 | 122 |
| 98 | 249.2222 | -46.6575 | 249.2102 | -46.6734 | -44.8 | -46.3 | -46.7 | -40.9 | 12.961 | 13.989 | 64.6 | 207 |
| 99 | 249.7808 | -52.8606 | 249.7466 | -52.8602 | -33.4 | -44.2 | -33.8 | -40.8 | 13.648 | 13.977 | 74.4 | 271 |
| 100 | 251.3589 | 55.5901 | 251.3984 | 55.5814 | -57.2 | -19.0 | -59.9 | -21.8 | 10.730 | 12.817 | 86.3 | 111 |
| 101 | 252.6635 | -63.8999 | 252.6712 | -63.9020 | -37.9 | -85.5 | -42.3 | -78.5 | 11.158 | 12.892 | 14.3 | 122 |
| 102 | 253.9912 | -58.7124 | 253.9837 | -58.7024 | -51.5 | -45.3 | -42.9 | -44.8 | 12.728 | 13.888 | 38.7 | 339 |
| 103 | 254.5424 | -48.8960 | 254.5416 | -48.9075 | -12.1 | 48.1 | -10.3 | 47.5 | 10.619 | 12.210 | 41.4 | 182 |
| 104 | 256.5224 | -24.6581 | 256.5494 | -24.6573 | -43.3 | -43.7 | -40.4 | -49.0 | 11.751 | 13.867 | 88.3 | 88 |
| 105 | 259.6444 | -62.6001 | 259.6445 | -62.6027 | -66.3 | -115.0 | -58.1 | -112.5 | 9.187 | 11.261 | 9.3 | 179 |
| 106 | 261.3732 | 57.0206 | 261.3591 | 57.0150 | -15.7 | 46.0 | -14.2 | 44.3 | 12.722 | 12.790 | 34.2 | 234 |
| 107 | 264.4178 | -16.7687 | 264.4173 | -16.7908 | 45.6 | 52.8 | 45.4 | 49.7 | 12.772 | 13.876 | 79.5 | 181 |
| 108 | 267.7704 | -37.8793 | 267.7887 | -37.8779 | -26.2 | -42.6 | -26.2 | -45.4 | 12.403 | 12.531 | 52.4 | 84 |
| 109 | 270.5582 | -33.4210 | 270.5647 | -33.4342 | -27.1 | -74.9 | -29.4 | -71.5 | 13.517 | 13.674 | 51.5 | 158 |
| 110 | 272.9142 | -26.5171 | 272.9175 | -26.5163 | -31.1 | 49.1 | -33.7 | 54.9 | 13.077 | 13.391 | 11.2 | 76 |
| 111 | 274.0208 | -25.3238 | 274.0038 | -25.3296 | -41.6 | -60.9 | -41.9 | -57.5 | 13.842 | 13.985 | 59.0 | 249 |
| 112 | 277.1616 | -14.6359 | 277.1614 | -14.6549 | 14.7 | 55.9 | 15.5 | 48.5 | 12.757 | 13.716 | 68.2 | 181 |
| 113 | 281.8960 | -6.9224 | 281.9024 | -6.9258 | -44.0 | 72.8 | -38.0 | 71.7 | 12.733 | 13.521 | 25.7 | 118 |
| 114 | 282.0128 | -7.1885 | 282.0112 | -7.1816 | -12.9 | 56.1 | -12.2 | 51.6 | 11.014 | 13.010 | 25.6 | 347 |
| 115 | 290.8635 | -13.7178 | 290.8796 | -13.7354 | -65.5 | -51.8 | -62.2 | -50.3 | 10.482 | 13.577 | 85.1 | 138 |
| 116 | 293.2054 | -3.6201 | 293.1888 | -3.6024 | -8.2 | 63.4 | -7.6 | 69.2 | 12.856 | 13.948 | 87.1 | 317 |
| 117 | 305.0078 | 62.2981 | 305.0105 | 62.2887 | 19.1 | 40.4 | 20.7 | 43.1 | 10.938 | 11.803 | 34.1 | 172 |
| 118 | 305.9958 | -66.3853 | 306.0034 | -66.3840 | -30.3 | -96.3 | -28.7 | -94.1 | 11.898 | 13.252 | 12.1 | 67 |
| 119 | 310.8831 | 52.2766 | 310.8795 | 52.2790 | 83.6 | 38.9 | 80.6 | 40.2 | 9.369 | 10.998 | 35.7 | 133 |
| 120 | 324.6577 | 56.6790 | 324.6642 | 56.6777 | 71.9 | -25.0 | 72.4 | -26.8 | 10.813 | 13.612 | 20.4 | 355 |

Fig. 63 - Candidate Common Proper Motion Pairs in UCAC4 – Part 3

In the interest of both fairness and completeness I am happy to acknowledge that UCAC4 has been a valuable resource for astronomical data-miners. Most of the problems that have arisen have had poor communication rather than poor quality data as the primary cause.

Enthusiasts will put up with a great deal if decisions are explained to them. "We are thinking of doing A, B and C because of D, E and F." is almost always an acceptable way to behave. However making unannounced changes, especially changes that are to be applied retrospectively, is almost always a public relations disaster.

Read my recent book (January 2014) to find out more.

Identifying Common Proper Motion Binary Star Systems
by Martin Nicholson
The book is available on Amazon (www.amazon.com)

Data mining using the recently published 1500 square degree near infrared proper motion catalogue from the UKIDSS Large Area Survey has yielded 1220 common proper motion binary star systems of which 1129 appear to be new discoveries. Each component of a common proper-motion pair can be considered to be at the same distance from the observer, of the same age and subject to the same degree of reddening. These pairs are an interesting area for both amateur and professional astronomers to research because they do not fall into either of the extensively studied groups of orbiting binary stars or open clusters. Only a small proportion of the astronomical data mining and astronomical imaging projects I carried out over the last 20 years went through the lengthy – and sometimes controversial – process of third-party publication. Some of the rest of my work has appeared on assorted web sites or within specialist society "news groups" but much of it remains unpublished. The long-established scientific principle "first to publish gets the credit" means that without publication taking place it is almost as if the work had never been done! "Research Topics for Amateur Astronomers" will be a multi-volume series containing two types of material. Research Notes are intended to share discoveries, ideas or techniques of interest to astronomers. Articles equate to a traditional peer reviewed article.

## BY THE SAME AUTHOR

All are available from Amazon.com and from Amazon.co.uk


1800 new double stars for amateur observers

3600 celestial asterisms for amateur astronomers

Discover your own variable star

Identifying Common Proper Motion Binary Star Systems

Identifying Identical Twin Star Systems from the SDSS Data Release 10

www.ingramcontent.com/pod-product-compliance
Lightning Source LLC
Chambersburg PA
CBHW081910170526
45167CB00007B/3231